# WORLD POPULATION
# ENERGY
# and
# CLIMATE CHANGE

**PERCY A PAYNE**

# DEDICATION

To Kailey, Addison, Olivia, Samuel, Kristen and Kent for whom I undertook the effort

To Anne, for her moral support and understanding during periods of doubt

# ABOUT THE AUTHOR

Percy A Payne obtained a Bachelor of Science degree in Mechanical Engineering and a Master of Science degree in Petroleum Engineering from Mississippi State University. He has studied and worked in the Oil and Gas industry for more than fifty years. He held Engineering, Research and Development and Senior Management positions. He participated in and provided management overview of technology developments including offshore deep water, arctic, ultra-deep and high-pressure wells, horizontal and fracking completions, thermal, and CO2 tertiary drilling and production methods.

He developed an understanding of energy's impact on economic activity and quality of life. He also gained an appreciation of the dominant position held by fossil fuels in world energy consumption and potential impacts on climate change of CO2 emissions from burning fossil fuels.

He envisions a dilemma involving energy supplies for future generations and the impact of CO2 emissions on the environment. Fossil fuels are a finite resource. Production levels are increasing, but, at some point, they will peak and begin to decline. As fossil fuels production levels grow in the short-term, CO2 emissions will also increase. A benefit of the production peak will be decreasing CO2 emissions; however, other sources of energy will have to be developed or energy consumption will decline negatively affecting world economies.

He is a father and grandfather. He expresses a heart-felt concern as to how this transition is managed and future lifestyles will be impacted.

He began a self-study program to address these issues. Significant energy options are available, but he believes that there will have to be a major transformation away from conventional petroleum and CO2 emitting fuels.

This will be driven by potential supply disruptions of petroleum and natural gas due to economic/political factors, shortages of petroleum and natural gas as the world reaches peak production capacity and potential impacts of CO2 emissions on the environment as they are better understood and quantified.

Due to the immensity of this undertaking- focused planning for future energy should begin now. In the near-time consumption of fossil fuels will continue to grow but utilization of alternate energy sources should be increased to halt growth and eventually reduce reliance on fossil fuels. He concludes that future lifestyles will depend on how this energy transformation is managed.

This writing represents the results of his study. It is intended as a guide for others who share a marked interest in this subject. The purpose is to provide background information to aid an understanding of evolving energy issues enabling readers to actively participate in discussions of energy options.

# TABLE OF CONTENTS

# LIST OF FIGURES

## PREFACE

Archeological evidence indicates modern humans evolved about 300,000 years ago. World population had grown to about 200 million by the beginning of the modern calendar 2000 years ago. By 1865 it had grown to 1.3 Billion. World population is 7.7 billion in 2019.

The rapidly expanding world population the last 150 years and an insatiable desire to improve living standards has resulted in exponential growth in energy consumption.

Energy powers automobiles, airplanes, buses, trucks, tractors, ships, and factories. Energy lights, heats, and cools homes, schools, offices, restaurants, shops, factories, government buildings, and hospitals. In effect, energy supports almost every facet of daily lives.

With the exceptions of the 1973 crude embargo by OPEC (Organization for Petroleum Exporting Countries) and disruptions due to tensions in the Middle East in the late 1980's, there have been abundant and inexpensive supplies of energy. The world has taken energy for granted, not recognizing the dependency on this critical commodity.

Data are presented for the total world as well as by major economic regions to show interdependence of energy production, consumption and trade.

Data were derived from numerous public sources in varied units. Energy conversions to common units were made for comparison in many cases. Melding sources and energy conversions resulted in minor anomalies in some cases but not sufficient to effect overall trends and conclusions. Interpretations of the data represent views of the author.

Conventional fossil fuels such as oil, coal, and natural gas supplied approximately 83 per cent of the energy consumed in 2015. These fuels are a major source of increasing CO2 levels in the atmosphere; a likely contributor to global warming.

Significant energy options are available but at some point, there will have to be a transformation in energy supplies and usage. It will be driven by short term disruptions in supplies of petroleum and natural gas due to economic/political factors; shortages of petroleum and natural gas after production reaches peaks and begin to decline; or reduction in fossil fuel consumption due to concern of impacts of CO2 emissions into the environment as they are better understood and quantified.

Great technological challenges will arise as a result of this transformation. If they are met, economic opportunities will be fostered. If not, the very standard of living the world has become accustomed to will be threatened.

The transformation has already begun and is the third in energy usage. The first was from wood to coal. The second was from coal to crude oil and natural gas. The third will replace petroleum and natural gas with renewables as the primary source of the world's energy

Positive forces drove the first two transformations: first, coal is a more convenient and versatile fuel than wood and second, crude oil and natural gas are preferred over coal. However, the current transformation will be more problematic. The need to replace petroleum and natural gas with fuels that do not contribute to global warming will be expensive and technically complex.

The purpose of this document is to provide a basic understanding of energy; its uses, its sources, how it is converted from one form to another, its estimated supply, its regulation and its potential impact on climate change. Given this background, readers will have the opportunity to follow developing issues and actively participate in energy-option discussions.

The material is presented so it can be understood without in-depth knowledge of energy's science and technology. Only fundamental scientific principles and basic technology concepts are discussed.

Two sections; SOURCES AND NATURE OF ENERGY and USE and CONVERSION OF ENERGY contain background information on energy. They are helpful for the total energy picture but not required to understand major energy issues. Accordingly, they may be skipped.

This writing does not focus on the efforts of those involved in finding, producing, converting, and supplying energy, although the citizens of the world are truly indebted to them.

# SUMMARY

World population was about 200 million at the start of the modern calendar over 2000 years ago. By 1865 it had grown to 1.3 billion, a six-fold increase. World population expanded rapidly the last 150 years with no end of growth in sight as shown in Figure 1. It is approximately 7.7 billion in 2019.

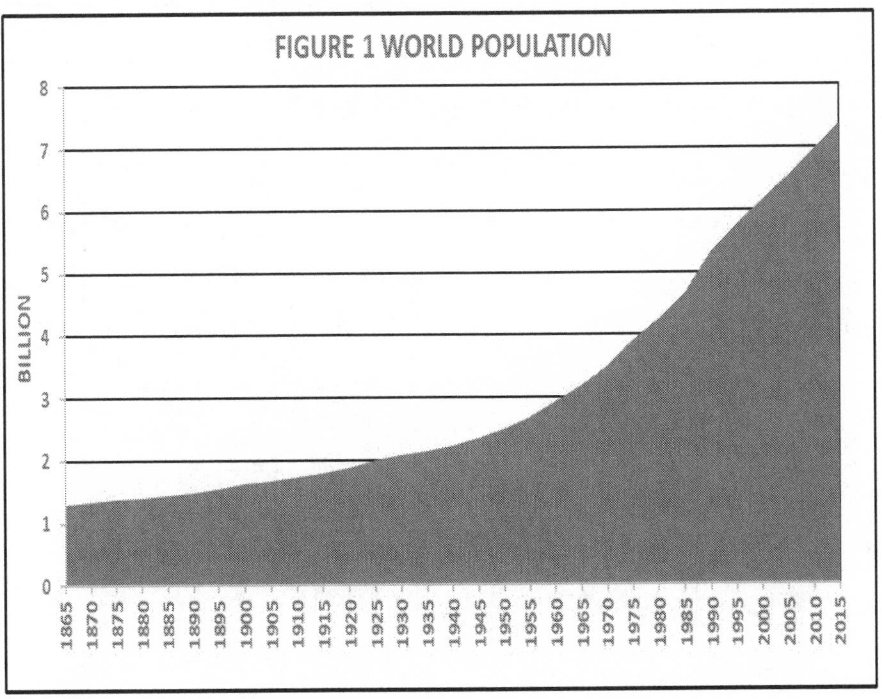


The expanding world population has many related issues that must be considered when discussing the well-being of planet earth. Energy production, consumption and potential effect on the environment are major ones and the focus of this document.

Expansive data on population, energy and climate change are available; this document attempts to organize data into useful information. Historical data is presented from 1865 to 2015. 1865 roughly corresponds with beginning of rapid population growth and the first oil well. Complete data for the varied charts is available for 2015. Later data is included in some discussions. Future trends are forecast to 2040 corresponding with Department of Energy, Energy Information Agency (eia) forecasts.

Growth in energy consumption per person has also expanded rapidly due to quality of life improvements in developed countries and modernization efforts in undeveloped regions. This is illustrated in Figure 2 which shows daily energy consumption per person. Gallons of oil equivalent per person per day is shown as the amount is easy to visualize.

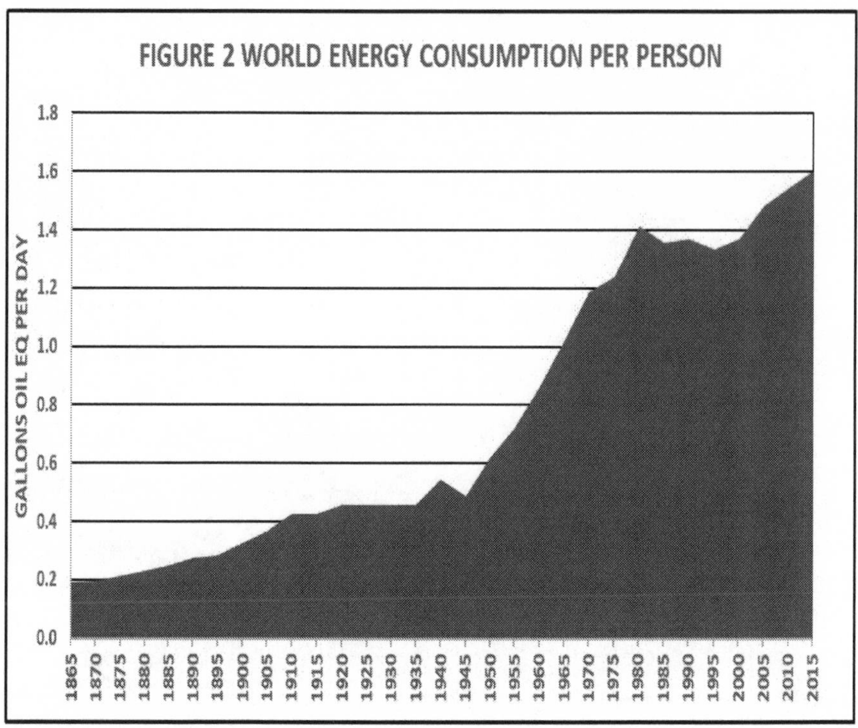

FIGURE 2 WORLD ENERGY CONSUMPTION PER PERSON

Energy consumption was .2 gallons of oil per day per person in 1860 compared to 1.6 gallons in 2015- an 8-fold increase.

Explosive growth in the world's usage of energy (particularly after World War II) occurred as a result of the two factors above Figure 3.

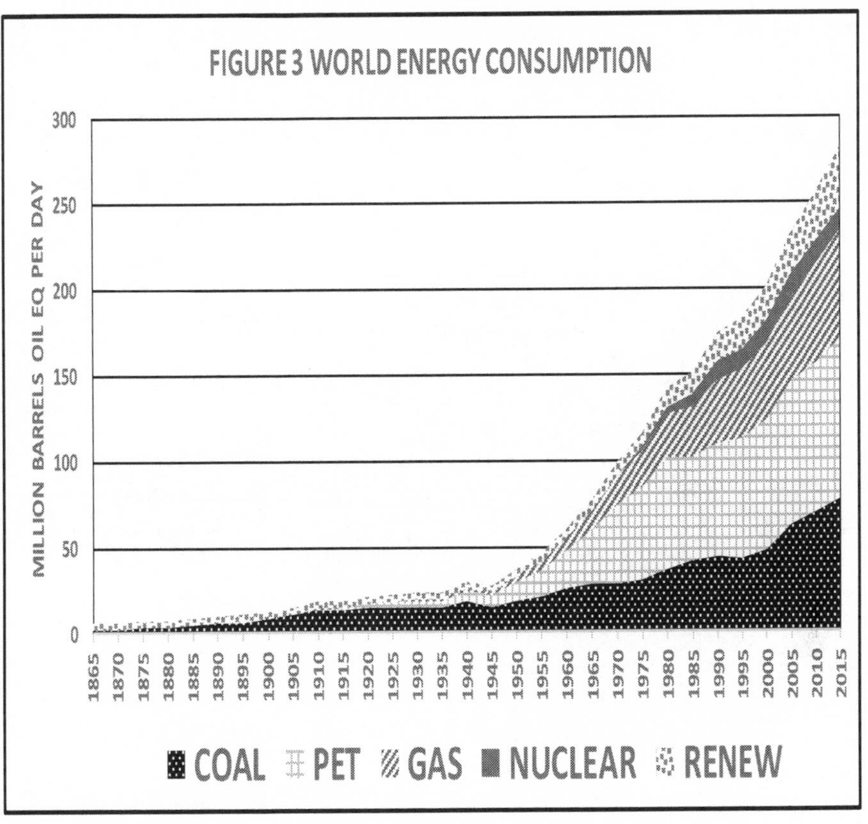


Usage in 1865 was 5.4 million barrels oil equivalent per day compared to 281 million in 2015, a 52-fold increase.

Figure 4 shows the same data presented as the percent that each fuel contributes to the total. Approximately 83 per cent is currently petroleum, natural gas, and coal (fossil fuels), which are non-renewable and finite in amount. These fuels are a major source of $CO_2$ emissions to the atmosphere, a contributor to climate change.

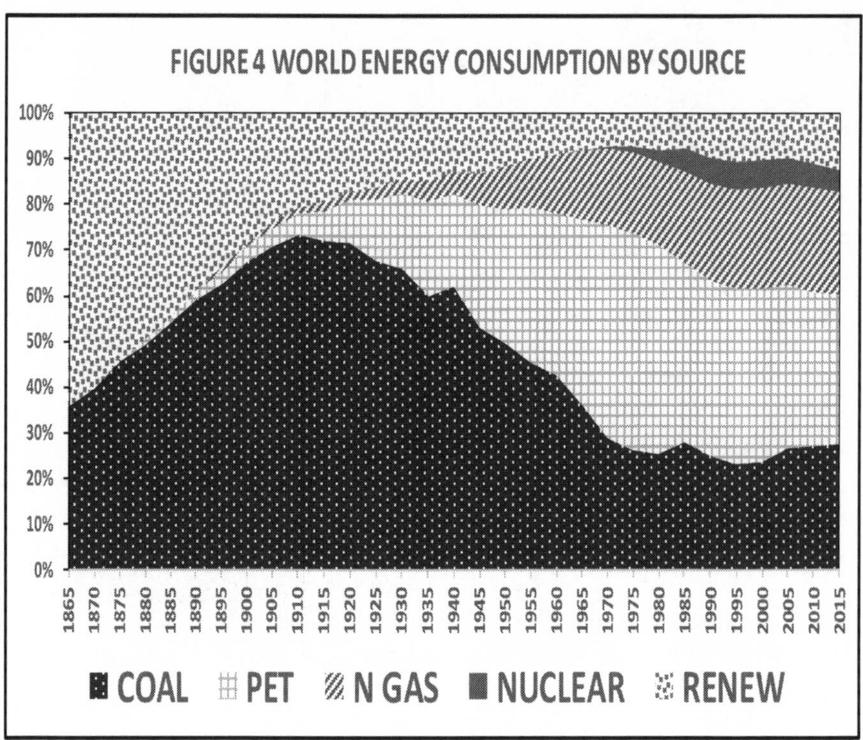


There has also been a rapid growth in CO2 emissions resulting from burning fossil fuels as shown in Figure 5.

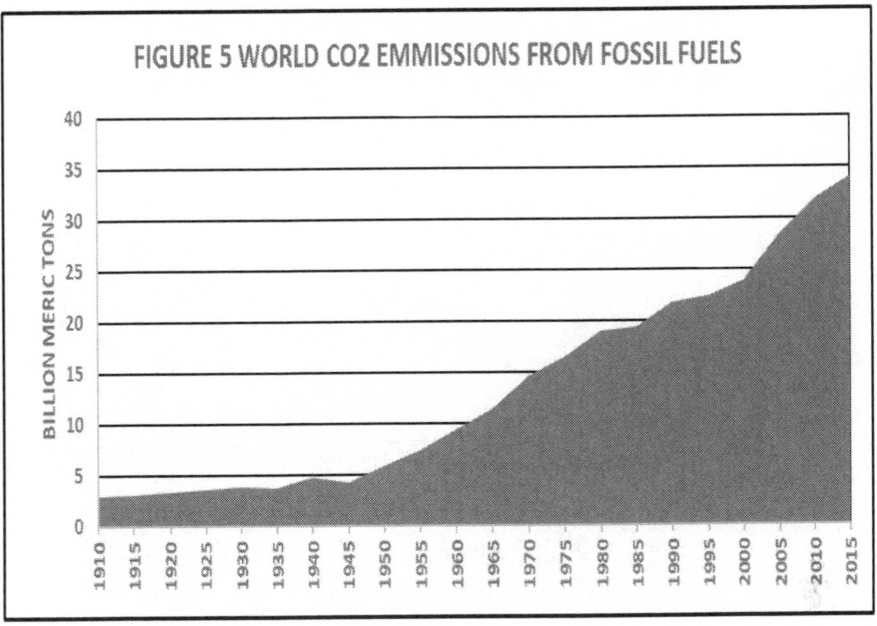


CO2 is known to be a greenhouse gas which contributes to climate change. An increasing concentration of CO2 in the atmosphere is reflected in Figure 6.

.

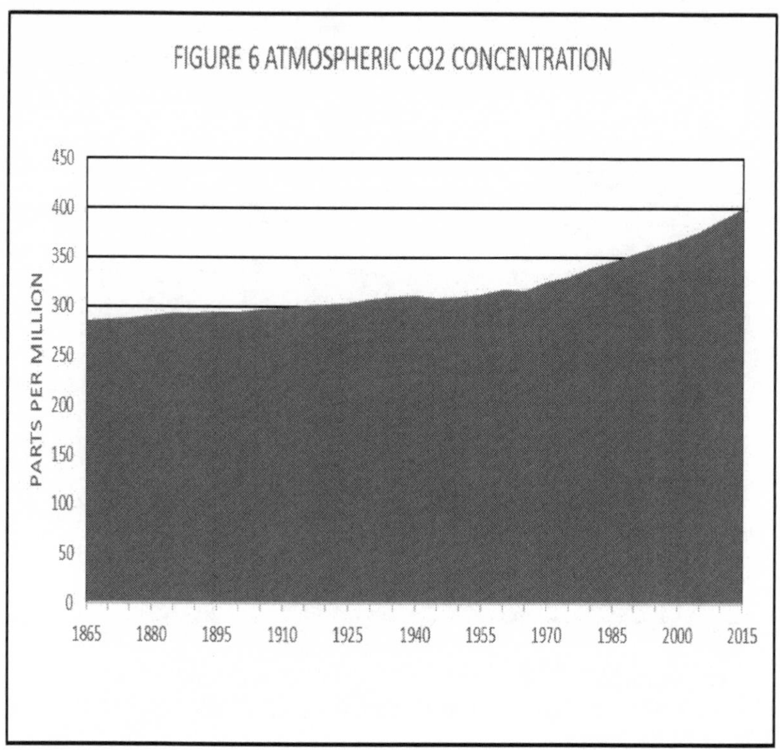


More importantly Figure 7 show historical estimates of surface temperature. Persistent rising global temperatures are indicated since 1980. An increase of .6 degrees Celsius or slightly less than one-degree Fahrenheit is indicated above the 1961-1990 average.

FIGURE 7

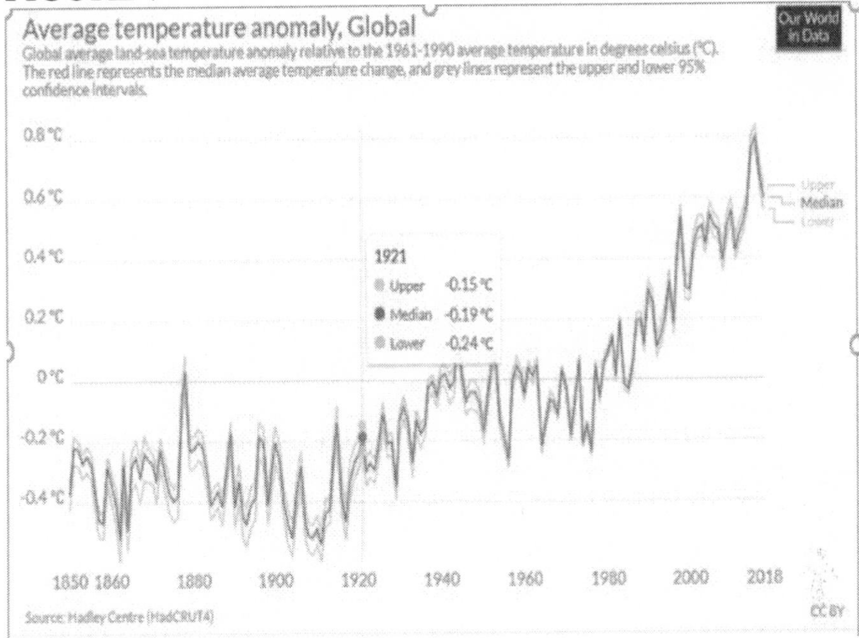


It is clear global warming is occurring. It started around 1900; long before fossil fuels were a factor. This suggests many factors (solar activity, ocean currents, volcanic activity, shifts is earths axis tilt etc) other than CO2 influence global temperature; however, since 1980 the rise is persistent and likely related to increased CO2 in the atmosphere due to increased burning of fossil fuels.

Impacts of global warming on climate change and economic activity are less understood. Both should be quantified before major adjustments in energy are made. Reductions in fossil fuels have major implications;

In 2018:

83 percent of the world's energy was supplied by fossil fuels

90 plus percent of transportation fuels came from petroleum

74 percent of the worlds primary energy was used for transportation and electricity

24 percent for transportation

50 percent for electricity generation

74 percent of electricity was generated with fossil fuels.

Transportation fuels in 2018 were petroleum based. Economic activity is heavily dependent on transportation. Replacement of petroleum-based transportation fuels is the most difficult issue if petroleum supplies fall short of demand.

Switching to electric vehicles for transportation before renewable sources of energy are available may not be helpful to the environment. It takes the same amount of energy to power the same weight vehicle. Electricity replaces petroleum as the power source. Currently 74 percent of this electricity is generated by fossil fuels. Electricity generating plants are more efficient than internal combustion engines but transmission and storage losses offset this. Further, coal, which has the most harmful emissions, generates 38 percent of the world's electricity today.

Also, renewable and nuclear sources of energy produce electricity which will be transported over the electrical grid. As output from these sources increase the grid will have be to upgraded. Upgrades should be in place beforehand.

Significant supplies of energy exist for the near term and future supplies can be managed, but several requirements must be met: (1) Wise choices must be made to reduce reliance on petroleum and natural gas if economic/political factors disrupt supplies in the near term and/or production peaks are reached and (2) climate change and potential effects must be better understood and quantified..

Alternate energy choices will be complex since options are many, varied, and technically challenging. Further, these choices must be made in the context of climate change.

Coal, which supplied 27.5 percent of the worlds energy in 2015 has sufficient reserves to offset declines in petroleum and natural gas production but its use will only increase CO2 emissions and contribute negatively to climate change.

Nuclear energy has the potential to increase significantly and does not emit any harmful emissions to the atmosphere but several issues impede development. Safety of nuclear plant operations and storage of spent nuclear fuel are the most controversial.

Renewable sources of energy such as biomass, wind, hydro, and solar must supply increasing amounts of energy in the future.

As a result of the changing mix of fossil fuels and increasing alternate sources of energy, massive energy infrastructure transformations will be required.

These transformations have worldwide implications. The intensity of the transformation of energy utilization and sources will grow as peaks in world petroleum and natural gas production approach and/or impacts of climate change are better understood.

The author is concerned, but not dismayed. Scientists, engineers and innovators developed the machines that propelled our standard of living to what it is today. Currently, these machines are fueled primarily by fossil fuels. This author feels that, given time and proper technical/social/political choices, this same inventive curiosity will develop required new energy sources and machines.

A lot of data is presented. Energy data is presented in barrels of oil equivalent (BOE) as often as possible to make comparisons easier. To facilitate discussion of world energy, eleven country groupings with similar energy and socio-economic patterns are used. The groupings are: Canada; United States; Venezuela; Mexico, Central and South America (excluding Venezuela); Europe; Middle East; CIS (which includes Russia); Africa; China; India; and Asia Pacific excluding China and India. Canada and Venezuela are included separately because they hold large amounts of heavy oil reserves.

The purpose of this document is to provide information and data so the reader can understand issues as they develop and actively participate in discussing energy solutions.

•

## IMPORTANCE OF ENERGY

Energy fuels the engine that drives economic activity. Energy powers automobiles, airplanes, buses, trucks, tractors, ships, and factories. Energy lights, heats, and cools homes, schools, offices, restaurants, shops, factories, government buildings, and hospitals. In effect, energy supports almost every facet of daily lives.

The economic importance of energy can be seen by examining the relationship between economic activity and energy consumption. Gross Domestic Product (GDP), the total of all goods and services produced by a country is used to measure economic activity. Gallons of oil equivalent per year is used for energy as it is easy to visualize. Reliable statistics are kept for both measures. The effect of population size is removed by normalizing the data on a per capita basis. The table below compares GDP and energy consumption for selected countries with 2016 data.

| COUNTRY | GDP US$/PERSON | ENERGY CONSUMPTION GOEQ /PERSON | ENERGY EFFICIENCY US$GDP / GOEQ |
|---|---|---|---|
| United States | 57,804 | 1,992.7 | 29.01 |
| Australia | 52,587 | 1,604.5 | 32.78 |
| Germany | 42,481 | 1,118.7 | 37.97 |
| U K | 40,436 | 809.6 | 49.93 |
| Japan | 38,757 | 1,004.7 | 38.58 |
| World Ave. | 10,828 | 562.3 | 19.26 |
| Russia | 8,898 | 1,448.3 | 6.14 |
| Brazil | 8,633 | 435.1 | 19.84 |
| China | 7,996 | 655.4 | 12.20 |
| South Africa | 5,280 | 789.9 | 6.68 |
| Egypt | 3,474 | 238.8 | 14.55 |
| India | 1,717 | 186.6 | 9.20 |

GOEQ: Gallons Oil Equivalent

General trends should be representative of trends that would emerge if all countries were shown.

It is apparent countries that enjoy robust economic activity consume the most energy. They also generate the most economic activity per unit of energy consumed; i.e., they use energy more efficiently. This conclusively demonstrates that economic activity (how we live) is closely related to how much and how efficiently energy is consumed.

Another significant point in the above table is that countries with lower GDP per person than the world average contains a large portion of population. This suggests a huge pent up demand for energy demand as these countries strive to grow their economies. As a result, there will be increased pressures on future energy supplies in addition to pressures caused by the expanding world population.

The above relationships developed while energy was readily available at reasonable prices primarily based on fossil fuels.

Properly planned, the transformation to renewable sources of energy can be accomplished without major economic disruptions. New economic opportunities should more than offset negative adjustments.

If the transformation is made under crisis conditions due to poor planning or unexpected world events, the economic fallout could be significant.

# SOURCES AND NATURE OF ENERGY

Except for that which is derived from nuclear reactions, the source of all energy is the sun. Energy is generated by the sun directly through voltaic activity (Solar panels) and photosynthesis or indirectly capturing energy from weather driven by the sun and the earth's rotation.

All flora (plants) grow by photosynthesis of the sun's energy beginning the carbon cycle which is the first key to all life on planet earth and hence energy. Fauna (animals, fish and birds) eat the plants (herbivores) or other fauna (carnivores) to grow. Some of these plants can be converted into energy (Biomass). Some of these flora and fauna formed fossil fuels as they were buried and were subjected to increased temperatures and pressure.

Since photosynthesis and the carbon cycle are so important a brief review follows. The process starts when plants extract carbon from carbon dioxide (CO2) in the air and expel oxygen (O2) into the atmosphere by photosynthesis of sun rays augmented by water which plants absorb from soils and/or air. Carbon is stored in the plants. Animals, fish and birds eat plants, inhale O2 from the air, absorb some of the carbon and exhale CO2. The process starts all over again in living matter.

The result is that carbon makes up 18 percent of animal matter i.e. humans consist of 18% carbon. Plant life is more complex as there are so many varied types but trees on average contain about the same amount of carbon, 18 percent.

Carbon in dead plants and animals can be converted into hydrocarbons (CH) if subjected to heat and pressure under proper conditions. Hydrocarbons react with oxygen in air at high temperatures (burning) which gives off heat, water vapor and CO2. The process starts over again.

Weather is the second key to life on planet earth. All living things require water. Weather driven by energy from the sun and rotation of planet earth moves water around the globe. This movement starts by water vapor being absorbed into the atmosphere. Energy from the sun heats air close to the earth's surface where air pressure is highest. Warm high-pressure air holds more water vapor so when it is in contact with water (i.e. over oceans) water vapor is absorbed into the air. Air containing more water vapor is lighter than dry air, so it begins to rise. As it rises it gets colder and pressure drops so it gives up water vapor in the form of rain, sleet or snow. The water is returned to the ocean and the cycle starts anew. Wind is caused by mass air movements augmented by the process and rotation of the earth.

Energy relates to the capacity to perform work. Two broad classifications of energy are potential and kinetic. Potential energy has capacity to do work if ignited or activated. Examples are fossil fuels and batteries.

Kinetic energy is due to a body in motion. It has capacity to do work as energy can be extracted as it slows down. Water can be used to demonstrate examples of both potential and kinetic energy. Water behind a dam at a higher elevation has potential to do work if allowed to fall to a lower level. As water falls it has kinetic energy that can be captured by using a waterwheel or turbine to drive machines to perform work.

Many forms of energy exist under the broader classification of potential and kinetic. Examples are mechanical, chemical, electrical, atomic, and thermal. All forms of energy can be converted into other forms with appropriate processes. In conversion processes, heat, the most transient form of energy, is normally involved.

Most fossil fuels are converted into useful forms of energy starting with heat given off by combustion. Nuclear fuels generate heat as atoms split apart in controlled reactors. In 2015, 88% of energy consumption was fossil fuels and nuclear. Hence, the science of energy is the science of heat. This will change in the future as developing forms of renewable energy result in electricity being the form of energy captured.

To understand this in more detail, it is helpful to review the makeup of matter. All matter is made up of 118 currently recognized elements. The basic building block of each element is the atom. An atom consists of a nucleus and one or more electrons. Electrons orbit the nucleus, much like planets orbit the sun. The nucleus is made up of two parts, protons and neutrons, collectively referred to as nucleons.

Two or more atoms may combine to form molecules. For example, two atoms of hydrogen and one atom of oxygen combine to form one molecule of water. Chemical reactions occur when atoms are rearranged, but not altered, to form new molecules. For example, water molecules are formed from oxygen and hydrogen atoms without a change in the makeup of the hydrogen and oxygen atoms.

Combustion (burning) is a chemical reaction that releases heat in the process. Most fuels are compounds of hydrogen and carbon. When oxygen is introduced from air and the fuels ignited, heat is released, and major new compounds of water and carbon dioxide are formed. The atoms are not altered.

Atomic reactions occur when the makeup of an atom is changed, such as in the splitting apart the nucleus of uranium atoms to form new elements with different atomic structures. As this happens heat is generated.

All energy is initially obtained from primary sources. Fossil fuels and nuclear are primary sources of energy and contributed 83% of energy consumed in 2015. Most was converted to other secondary forms before use. The most notable source of secondary energy is electrical energy which is generated from primary sources. Hence, electrical energy is not a source of energy at all but a convenient and flexible form of secondary energy.

Primary sources of energy are grouped broadly as Fossil Fuels, Nuclear (Atomic), and Renewable. These groupings will be used in the discussion below.

## FOSSIL FUELS

Fossil fuels supplied 83 per cent of the world's energy in 2015. They were formed millions of years ago from plant and animal matter (biomass). Plants grew as a result of photosynthesis of the sun's energy. Animals grew by eating plants and/or other animals. Plant and animal matter from both land and sea were partially formed into fossil fuels due to being buried and subjected to increased temperature and pressure. They were buried due to erosion of highlands and are in "storage". They are extracted from the earth as needed. They are finite in amount so we are depleting the supply as they are consumed.

Fossil fuels consist of compounds made up of carbon (C) and hydrogen (H) atoms (commonly referred to as hydrocarbons). For example, natural gas-a simple fossil fuel, consists of one carbon atom and four hydrogen atoms ($CH_4$). Gasoline, a more complex fossil fuel, is made up of on the order of eight carbon atoms and a proportionate number of hydrogen atoms.

Energy is produced by a chemical reaction of hydrocarbons and oxygen. This reaction is commonly called burning and/or combustion. Air, which is 21 per cent oxygen, is the normal source of oxygen. Burning is initiated by raising the temperature sufficiently to start the chemical reaction. Once initiated, the reaction will continue until the fuel is consumed or the source of oxygen is shut off. The product of combustion of hydrocarbons is water ($H_2O$), carbon dioxide ($CO_2$), and heat. Combustion of natural gas is shown below;

$$CH_4 + 2O_2 \rightarrow CO_2 + 2H_2O \text{ (water)} + \text{HEAT}$$

Note in the above example that the molecules are rearranged, but the atoms are not altered. Also note the formation of CO2 in the burning process. CO2 is a greenhouse gas which contributes to climate change.

There are three broad families of hydrocarbon fossil fuels: crude oil, natural gas, and coal. Hydrocarbon compounds that are in a gaseous state at atmospheric conditions are commonly referred to as natural gas. Crude oil is normally found in a liquid or semi-liquid state. Coal is made up of hydrocarbon compounds that are normally found in a solid state.

Petroleum is sometimes used interchangeably with crude oil, but it normally refers to liquid hydrocarbons such as crude oil, natural gas liquids and refinery gains. On a world basis natural gas liquids and refinery gain volumes are about 10% of the volume of crude oil. Hence petroleum is about a 10% larger volume than crude oil alone. Some data sources present crude oil and petroleum interchangeably. Differences this may cause are not significant in the larger picture.

The most versatile of fossil fuels, crude oil, is considered by some to be the "noble" fuel. Easily transported and stored, it is refined to manufacture most of the world's transportation fuels such as gasoline, diesel fuel, aviation fuel, and bunker fuels for ships (also used in electrical generating plants). It also supplies most of the world's lubricants, such as motor oil and industrial grease.

The chemistry of hydrocarbons is complex and is further complicated by impurities that might be contained in the compounds. However, for energy purposes, it is only important to know that as hydrocarbons burn, heat is released which is then used to perform work or converted to other forms of energy. A significant by-product of burning hydrocarbons is CO2; a greenhouse gas. The amount of heat and CO2 released varies with the type of hydrocarbon.

## FORMATION OF HYDROCARBONS

Hydrocarbon compounds were formed from decaying plant and animal matter (biomass) buried in sedimentary basins. Sedimentary basins were low areas such as swamp regions, coastal areas, or those under the ocean where sediments consisting of biomass and soils/rocks were deposited. While it is likely that most of the biomass grew in or near the sedimentary basins, the soils/rocks that filled the basins were blown in by wind or washed in by rivers from highlands.

The core of the earth is very hot, causing a temperature gradient from the earth's surface increasing toward the center. As soils and biomass were buried by the continuous filling of the basins, they were subjected to ever-increasing temperatures due to the temperature gradient and pressure due to the weight of the sediment above. Depending on the makeup of the biomass and the temperature and pressure to which it was subjected, some of the biomass formed into hydrocarbon fuels as coal, oil, and natural gas. This is a simplified description of a very complex process. This progression took millions of years and for all practical purposes, the supply is not being recreated as it is used.

Coal was formed as areas supporting prolific plant matter were buried. Before the burial process began, layers of plant matter many feet thick covered the ground. As the burial process progressed, those layers were compressed and formed into coal "seams" from a few inches to several yards thick. The quality of coal found today depends on the burial and formation history and consists of three basic types: anthracite is the purest form and cleanest burning, bituminous is the next higher quality and lignite is the lowest quality and normally has more harmful byproducts in the combustion process. Coal is mined from underground seams either by stripping away the layers of earth above them or digging shafts and/or tunnels to them.

Crude oil and natural gas were formed in either a liquid or a gaseous state and were capable of flowing underground given a flow path. Porous strata and fractures in the earth provided those flow paths. When flow paths were blocked by layers of impermeable rock, oil and gas accumulated in underground "reservoirs."

Underground reservoirs consist of two types of rock: sandstone and limestone. Sandstone reservoirs began as deposits of sand that were buried as the basin filled. The sand deposits might have been a river sandbar, a beach, a delta, or simply an accumulation of sand that washed into a low area. There is porous space between sand grains. It is in this space that oil and gas are found. The amount of porous space depends on the makeup of the sand grains and the burial history of the rock.

Limestone reservoirs were formed in an ocean environment through a much more complex process. They were initiated by an accumulation of organic sea life remains, such as shells and reefs. Subjected to higher pressures and temperatures as the basins were buried, these remains were transformed into principally calcium carbonate rock. Porous spaces were either left behind during the burial and formation process or formed as portions of the rock were dissolved by underground water action. The quality of limestone reservoirs varies widely depending on the burial and formation history.

The quality of a reservoir is determined by the amount of storage space in the rock and the ease with which the oil and gas can flow from it. We drill into these reservoirs today to extract crude oil and natural gas. The science and technology involved in finding, drilling, and extracting the oil and gas is beyond the scope of this book but many writings are available for those interested.

Coal, oil, and natural gas as described above are considered "conventional fossil fuels," because extraction techniques are relatively inexpensive and well known.

# UNCONVENTIONAL FOSSIL FUELS

More biomass was formed into hydrocarbons than is convenient to mine and extract as discussed above. To obtain these, more complex and/or expensive extraction technology is required. These are referred to as "unconventional fossil fuels" and consist of four basic types:

1. Natural gas was formed concurrent with coal (coal seam gas) and remains trapped in the coal. For years, this was a hazard to coal mining. The potential is now being realized with the production of natural gas from coal seams prior to coal mining or as a stand-alone process. Natural gas can be produced from coal seams where coal mining cannot be justified making many more coal beds a valuable source of energy.
2. Tar sands and/or heavy oil deposits accumulated after crude oil migrated into shallow sands. The crude oil was formed at deeper depths but most of the lighter hydrocarbons evaporated or were acted on by bacteria, leaving a mixture of very heavy crude or tar and sand. The tar can be utilized after separating the sand by upgrading it with appropriate refining processes. Deposits too deep to mine are produced by introducing heat (primarily steam) to make the oil less viscous allowing it to flow to producing wells.
3. Shale oil deposits consist of fine sediments silt and clay containing a blend of kerogen. Kerogen is an immature crude, since its formation process was interrupted. Usually the fine sediments and biomass were deposited in deep freshwater lakes and buried. Before the crude matured, uplift occurred, and the formation process was halted. This kerogen can be processed to form low-quality crude oil after the silt is removed.
4. Tight sands and some types of shale contain conventional natural gas and crude oil trapped in formations offering restricted flow paths. Over geologic time, natural gas and crude oil migrated into them or were formed there directly. To produce the gas and crude oil, artificial flow paths must be induced by fracturing the earth and installing sand to provide a flow path to the well.

Unconventional fossil fuels represent significant future supplies of energy. Production is rapidly growing as technology is improving. Significant improvements have been made by combining horizontal drilling with fracking technology as illustrated below:

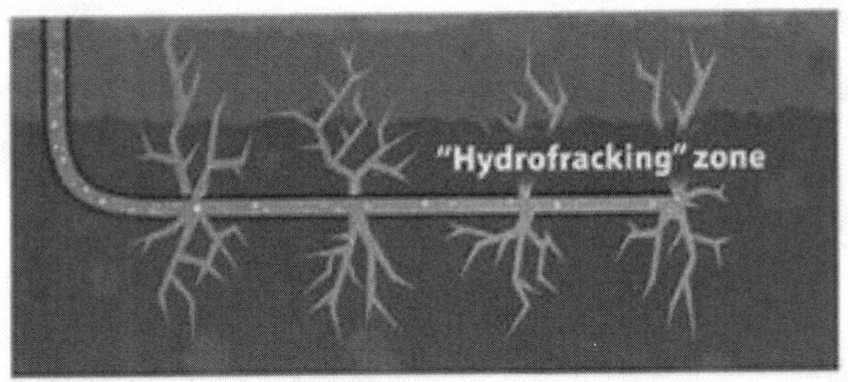


Increased production from horizontal wells with multiple fracture treatments has bought United States crude production in balance with consumption and allowed it to become a net exporter of natural gas. More importantly this technology has stabilized world oil and gas supplies at attractive prices.

FOSSIL FUELS TECHNOLOGY

Horizontal drilling with fracture treatments is just the latest in a long history of technology developments to find and produce fossil fuels. The first oil wells were dug by hand in or near surface oil seeps.

Geologists first used surface expressions to infer the shape of underground structures. Imaging methods using sound waves (induced seismic) were then used to map much deeper underground structures. As computer technology developed, the ability to see underground progressed to the point structures are mapped in 3 dimensions today. More detailed underground mapping technology is being developed as computer capacity increases. It is now possible to walk around in an underground virtual reservoir.

Engineering and operating procedures were concurrently developed to drill for and produce oil and gas found by geologists.

From simple wells onshore, the arena moved to deeper-higher pressure wells onshore, bottom supported platforms offshore, floating platforms in deep water and the arctic regions.

Petroleum engineers working with geologists developed methods to recover more and more oil and gas from known accumulations. Better mapping and computer simulations greatly aided this effort.

A significant side benefit of this effort is the use of CO2 to facilitate recovery of oil left behind after normal production. It is now proven technology with added potential to reduce CO2 emissions to the atmosphere.

Innovative thinking and experimental development enabled quantum steps in finding and producing fossil fuels. Similar breakthroughs are likely for nuclear and renewable sources of energy in the future.

## NUCLEAR ENERGY

Nuclear reactors supplied 4.5 per cent of the world's energy in 2015. Nuclear energy is derived as a result of the atomic structure of one or more elements being altered. The nucleus is split, changing one element into two or more and releasing heat; hence, the term nuclear energy. This is possible as a result of the atomic characteristics of certain elements. They are radioactive fissile elements, i.e., they are capable of fission or breaking apart.

The nuclear energy we use today is derived from these radioactive elements. A type of uranium is the only known naturally occurring fissile material. Like fossil fuels there is only a finite amount of this fuel available. The possibility exists to greatly increase the efficiency of nuclear reactors by "breeding" other fissile materials. Further, at least theoretically, there is the potential to obtain energy from a fusion process when two elements are "fused" with a net gain in energy.

Recall that all matter is made up of 118 elements, each with its unique atomic structure. The number of protons contained in the nucleus is used to classify elements (atomic number). Some elements exist with more than one atomic structure caused by different numbers of neutrons. These are called isotopes. Most elements are stable, and their atomic structures have not been altered throughout time. The oxygen we breathe today is the same oxygen our forefathers breathed. Iron we use today is the same iron that moved mankind from the Stone Age. The form may change, but the atoms making up the elements of iron and oxygen are not altered.

However, there are elements that can be induced to split and release energy. Such elements are said to be fissile, and the process of splitting is called fission. Fission is initiated in a fissile material by introducing free neutrons. The free neutrons react with nuclei of the fissile material, causing them to split forming new elements releasing excess neutrons and heat. The excess neutrons react with other nuclei to sustain the process. The process can be represented similar to the chemical equations as shown below:

U-235 + neutron → X + Y + HEAT + neutrons

The products X and Y are of interest. First, their form is not easily predicted and initially, they are not stable, i.e., they tend to be radioactive and thus contribute to disposal problems. However, they have significant upside potential to increase the output from nuclear reactors.

The uranium isotope U-235 is the only naturally occurring fissile material. Only about one in 140 atoms of natural uranium is U-235, only 0.7%. The bulk of natural uranium is U-238. Two other fissile materials are known, plutonium-239 and uranium-233. They are by-products of nuclear reactions i.e., they are bred. U-238 is the starting point which is not naturally fissile. However, it is the beginning point for the formation of plutonium with the "breeding process" continuing as shown below:

U-238 captures a neutron and becomes U-239;

U-238 + neutron → U-239

U-239 emits an electron and becomes Np**-239;

U-239 - electron → Np-239**

Np-239** emits an electron and becomes Pu-239*;

Np-239** - electron → Pu-239*

*Plutonium

**Neptunium

There are several significant positive and negative implications of the above process:

On the negative side plutonium is a very hazardous substance, it is difficult to dispose of plutonium nuclear waste and Plutonium can be used to make weapons of mass destruction.

On the positive side plutonium is fissile, fission of some of the produced plutonium occurs in the reactor, enhancing the energy produced by the reactor. Thus, reactors designed to produce and consume as much plutonium as possible can increase efficiency.

Finally, a "breeder" reactor can be designed to produce more fissile material in the form of plutonium than it consumes in the form of U-235. This plutonium can be extracted to use in other reactors. Recall only about .7 per cent of natural uranium is U-235 with the remainder being U-238. The possibility exists then to increase the yield of natural uranium resources by a hundred-fold.

Therefore, the amount of nuclear energy available from fission processes is highly dependent on the technology employed.

It is also possible to obtain energy from the fusion of two nuclei. Fusion is the process of joining two nuclei into one. The hydrogen bomb works on this principle. In such a thermonuclear device temperatures and pressures are extremely high and the reaction occurs very rapidly, making it difficult to harness released energy.

Theoretically it is possible, however, and a lot of work has been done toward this end. To date, success has not been reported, even in the laboratory.

## RENEWABLE

Renewable energy supplied 12.4% of the world's energy supply in 2015. Renewable energy is derived by direct conversion of the sun's energy or indirectly by converting natural forces fueled by the sun into energy. An example for direct conversion is obtaining electricity from solar panels by photovoltaics. An example of indirect conversion is obtaining electricity by running water from higher elevations through a turbine. The water was placed at higher elevations by weather events fueled by the sun.

Renewable energy will always be available to us. The amount that we harness is dependent on our ingenuity and the economics of converting it into useful forms. The total amount of renewable energy available to us is monumental. The problem is that there are limits on how much we can practically capture.

## SOLAR

Solar energy is obtained directly from the sun. In its simplest form, we may stand in the sunlight to break the chill on a bright winter day. We may also use a photoelectric cell to power a small remote electronic instrument. These two examples illustrate two basic physics concepts involved in solar energy, radiant heat and photovoltaic.

Both radiant heat and photovoltaic are caused by waves emitting from the sun. The underlying laws of physics, dealing with electromagnetic waves and photons, explain the process. Both travel at the speed of light, and can travel through a vacuum, allowing energy to flow from the sun to earth. Another key factor is that this energy is emitted in all directions uniformly by the sun and travels in a straight line until it is deflected or absorbed. The diagram explains several things:

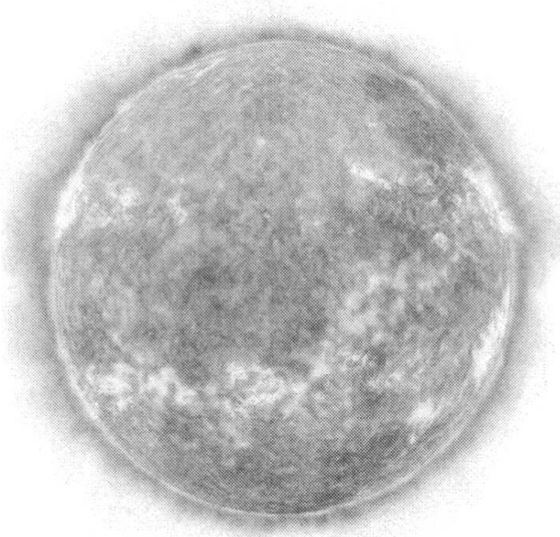

Sun Diameter: 864,600 miles Earth Diameter: 7,918 miles
Sun Temp.: 10,000 degrees F Earth TemP.: 57 degrees F
Distance between earth and Sun: Average about 92 million miles

The sun is emitting energy at a very high intensity due to its high temperature. This energy is emitted uniformly from its total global surface into space. The earth receives only a tiny portion of the energy emitted by the sun (roughly one billionth) since only rays emitting from the sun that intersect the earth 92 million miles away impact the earth on one side at any given time. The earth is also emitting energy into outer space over a much larger area than the sun energizes. However, the energy intensity of energy emitted by earth into space is much less than that received from the sun since the earth's temperature is much less. The result is the earth has reached a state of equilibrium whereby the amount of energy it receives from the sun is roughly equal to the amount of energy it emits to outer space.

We have observed that the sun's energy travels at the speed of light; and, like light, it can travel through a vacuum. Further, it behaves like light when it reaches a body of matter. That is, it is either reflected or absorbed. With this in mind, the earth's atmosphere consisting of dust particles and minute water droplets of moisture must be considered. Individually, these components tend to reflect the radiant rays, resulting in "scattering" of the rays and/or reflecting them back into space. Thus, we do not feel the sunlight on a cloudy day. In addition, the distance the rays must travel through the atmosphere affects the scatter. These factors vary between time of day, weather conditions and seasons. This effect should be considered when estimating the amount of radiant energy, we receive from the sun at a given spot on the earth.

The above illustrates difficulties with solar energy. It is only available during daylight hours. Its intensities are less in the morning and afternoon and in the winter months. Also, atmospheric conditions such as clouds, dust, and smog can significantly reduce amounts of energy received. However; much more energy than we consume does reach earth despite these difficulties. The problem is capturing, converting into usable forms, and storing it for later use.

Nonetheless, a great deal of solar energy can be captured effectively by taking advantage of simple concepts. For instance, a house built with windows facing south is positioned to capture the maximum amount of solar energy available in the winter. The walls and floors can be configured and constructed in such a way to absorb heat during the day and release it during the evening. Solar panels, with water circulating through them, can be placed on the roof of a house. During the day, the water is heated by solar energy. It could be used directly during the day or stored in large storage tanks for later use as hot water or heat for the house. In many parts of the United States, it is possible to provide almost 100 per cent of needed household heat.

The above systems, while useful, are limited in their utilization since the heat captured never results in high temperatures. For efficient conversion, relatively high temperatures are needed. Large solar collectors can accomplish this. One type of largescale solar collector consists of an array of mirrors covering several acres that concentrate the radiant energy on a given point. At that point, water can be converted to steam to drive turbines to generate electricity.

Utilizing photovoltaics, solar radiation can be converted directly to electricity by panels of semiconductors. This technology is rapidly being improved and is currently competitive with other energy forms. A complete system consists of solar panels and battery backup. When the sun is shining, the panels provide power to the application and/or to charge the batteries. During periods of low or no sunlight, the batteries provide power. With large arrays of panels, enough electricity can be generated for use in more conventional applications.

Development of battery technology has allowed efficient storage of electrical technology in cars and trucks for the transportation sector.

The above systems, designed to capture and utilize solar energy effectively, are technically complex but technology is rapidly advancing.

## HYDRO

Hydro energy is a secondary form of solar energy. It results from the vaporization of water by solar energy being subsequently deposited at higher elevations in the form of rain, sleet, or snow. Energy can be extracted as it is released to a lower elevation.

The ability of air to hold water vapor is a function of both atmospheric pressure and temperature. Air holds more water vapor as pressure and temperature increase. Atmospheric pressure and temperature are greatest at sea level and decrease at higher elevations. Consequently, the ability of air to hold water vapor is greatest at sea level and decreases at higher elevations.

Further, air containing water vapor is lighter than pure air and the density of air decreases as temperature increases. As energy from the sun warms air over water, it absorbs water vapor and the warm vapor-laden air rises. As these air masses rise to higher elevations or mix with cooler air masses they are cooled, the pressure is reduced, and the water vapor condenses, falling to the earth as water in the form of rain, snow, or sleet.

When water flows from higher elevations it represents potential energy, since energy can be extracted as it flows back to sea level. This energy has been used throughout history. At first, the natural current of rivers may have been used to float things downstream. Waterwheels (still used today) were an early mechanical method of capturing this energy.

Today, most hydro energy is converted into electricity with the use of dams and water turbines. Dams build up the potential energy of the captured water, while water turbines drive generators to convert the hydro energy into electricity. This is the origin of the familiar term, hydroelectricity. A hydroelectric complex works by controlled flow of water from a dam through water turbines which drive generators to generate electricity.

There are two important factors in the process: the height of the water behind the dam and the rate at which it can be released through the turbines. More energy can be extracted as the water level behind the dam rises and the volume that is released increases. Both are controlled by nature, and the designers must match the design to what nature provides.

The surrounding terrain determines the height of the water behind the dam. The volume of water in the reservoir and the frequency and the rate of replenishment by rain or snow determine the rate of flow. A well-designed complex would allow a near-constant flow of water through the turbines, generating a constant flow of electricity, with a rising and falling level of water in the reservoir balancing the seasonal variations in the amount of rain or melting snow. This leads to a very significant parallel benefit of complex-flood control.

. Large amounts of electrical energy are generated from such complexes. For example, Arizona's Hoover Dam complex alone delivers on average 4.2 billion-kilowatt hours electricity per year. It would take 13,000 barrels of oil per day to generate as much electricity. It is only the 79th largest complex in the world.

## WIND

Wind energy results from solar energy. As the sun heats air masses depending on the time of day or night, the season, and/or other climatic conditions, they tend to contract or expand and rise or fall. The result is constantly moving air masses or wind. Wind is more predictable in some areas such as mountain passes and along shores but less so in other areas. The velocity of wind ranges from gentle breezes to violent storms such as hurricanes and tornadoes.

Mankind has utilized wind energy for hundreds of years. Windmills and sails for boats and ships, the first common conductors, are still used throughout the world.

Like hydro, recent applications employ the wind to drive generators to produce electricity. There are many different designs attempting to obtain optimum performance, but they are simple in concept. Large windmills which drive electrical generators are now a familiar sight around the world.

The pitfall in wind energy is the unpredictability of the wind's timing and intensity. For this reason, most applications to date are in mountain passes and coastal regions where the wind is more constant. Total potential is great in other areas, just less efficient and more complex to manage.

Wind energy captured and supplied into electrical grids is increasing as technology improves. Improved battery technology to store electrical power for use when the wind is not blowing will further increase utility of wind energy.

## TIDAL

Tidal energy, the result of potential energy from water pressure, is a form of hydro energy.

Tides result from gravitational pull of the moon and sun on the seas and oceans as earth rotates. The difference in water level between a low and high tide can be quite large in some places--as much as forty-eight feet in the Bay of Fundy, Canada.

Placing a barrier across the mouth of a bay and controlling the flow through turbines placed in the barrier can capture tidal energy. High volume low head turbines driving electrical generators can be used. Beginning at low tide, as the tide rises, water pressure will build up on the seaward side. This pressure will drive the turbines as water is forced to flow through them. The opposite results, with pressure building up behind the barrier, as the tide falls and turbines are driven to allow water to flow out in a controlled fashion.

Massive structures are required to control the tidal flow. In addition, they must be capable of withstanding storms that might pass through. This makes the initial cost of tidal very high. Further, there are only two legitimate potential sites in the United States, the

Passamaquiddy-Cobscook Bay in Maine and the Cook Inlet in Alaska with a combined potential of only 3.6 Giga watts. Due to this low potential, tidal energy is not discussed in the chapter on resources.

There are other potential ways to capture energy from the oceans. Among these are wave action and deep thermal currents, although neither will likely make a significant contribution in the next 25 years. Therefore, they will not be discussed further.

## BIOMASS

Energy from biomass has three significant advantages over other forms of renewable energy:

1) It is a form of renewable energy that can be converted to liquid fuels which are easy to store and transport. Also, liquid fuels from biomass can be used in place of petroleum for transportation fuels.
2) It is more predictable than some forms of renewable energy. Planting and harvesting crops can be planned
3) It can be converted from existing waste products.

Energy from biomass, as the name implies, involves the procurement of energy from the biological forces of nature. Plant matter grows through photosynthesis of the sun's energy. Plant matter can be used as fuel directly, or, through other biological or chemical processes acting on plant matter, other fuels such as natural gas and a form of petroleum can be produced.

Wood from the forests is by far the most common current use of biomass for fuel. Dung (dried animal waste) and straw from grasses are other commonly used fuels in parts of the world where there are no forests.

Waste products are burned in many uses. The forest products industry has made great strides in burning bark, chips, and other waste products for fuel.

In other places, methane gas, generated by bacteria and enzymes ingesting waste, is recovered from sewers. Methane is also generated and recovered from landfills in similar fashions.

In total, biomass accounts for only about 5 per cent of the energy used in the world today. Growth will come from planned farming and conversion systems. Plants for these systems can be cultivated both on land and in the ocean.

The plants can be processed in two primary ways: through bacterial and enzyme action producing methane gas, as discussed above, and through fermenting processes which produce ethanol, a form of alcohol. More sophisticated processes can be used to produce other fuels similar to gasoline and diesel fuel.

Finally, some plants produce fuels directly. For example, some vegetable oils can be refined into diesel fuels rather easily. Some types of algae have the same potential.

The above is encouraging. The potential exists to grow energy needs and not depend on crude oil for transportation fuels. The potential is real, but a few words of caution are in order.

On a mass-produced basis, a favorable energy balance in the total process must be obtained. For example; if fertilizers derived from fossil fuels are used to grow biomass and fossil fuels are used to provide heat for the conversion process, a sufficient increase in energy may not be obtained. In this case fossil fuels should be used directly.

## GEOTHERMAL

Recall that the earth has an increasing temperature gradient with depth. This gradient is not uniform around the globe and in some places, there are hot spots where the temperature increases quite rapidly. If there are porous rocks containing water at such places, the potential exists for geothermal energy.

The geysers in Yellowstone Park are examples of such a situation where the water has a path to the surface. As the water is heated underground, it begins to form steam. As the pressure builds up, some of the water and steam are vented to the surface and form the geysers.

To capture geothermal energy, a well is drilled into the heated porous rock. The pressure is released, allowing the hot water to flow to the surface where it vaporizes into steam. The steam is run through a turbine driving a generator to generate electricity. If the water is recharged from the surface, then the system is capable of renewing itself.

Even if the water is not hot enough to produce steam, useful energy can be obtained by using the hot water to heat buildings, etc.

As explained, there are many varied sources of renewable energy. Further, only the major ones have been discussed. This raises the question why renewable sources supplied only about 12.4 per cent of the world's energy in 2015.

The reason lies in the convenience and historic low cost of fossil fuels versus the unpredictability and relatively more expensive cost of renewable energy.

## USE AND CONVERSION OF ENERGY

Energy is routinely consumed without thinking of how dependent daily lives are on readily available and inexpensive energy. Cars, trucks, buses, ships and airplanes are fueled by gasoline, diesel fuel and jet fuel; homes, schools, hospitals, factories; restaurants and offices are lighted, cooled and heated by heating oil, natural gas, coal and/or electricity; etc. In the preceding section primary sources of energy were discussed. Before energy is used as above, it often must be converted from its source form to other forms. To aid in understanding the total energy picture, this section is a review of how energy is converted from basic sources to useful forms.

Three basic principles of physics are central to the conversion and use of energy; The First Law of Thermodynamics, A consequence of the Second Law of Thermodynamics; and The Second Law of Motion.

The First Law of Thermodynamics simply states, "When a closed system executes a cycle, work output realized is proportional to net heat input." This law establishes the relationship between work and heat (heat is a form of energy) and explains why most primary sources of energy are used to generate heat that energizes engines to perform useful work.

A consequence of the Second Law of Thermodynamics is that the efficiency of a thermal engine can be no greater than the following:

$$\text{Maximum efficiency} = \frac{To-Ta}{To}$$

Where:
To = Operating temperature of the system
Ta = Ambient temperature surrounding the system.

The temperature scale in the above is degrees Rankine, which is degrees Fahrenheit plus 460. The Rankine scale is sometimes referred to as the "absolute scale," since at 0 degrees Rankine, all motion would theoretically stop. A few quick calculations show efficiency increases as To increases.

For example, assume an ambient temperature of 70 degrees Fahrenheit or 530 degrees Rankine. For various operating temperatures, the following efficiencies would result:

| Degrees F | Degrees R | Eff. % |
|---|---|---|
| 100 | 560 | 5.3 |
| 500 | 960 | 44.8 |
| 1,000 | 1,460 | 63.7 |
| 1,500 | 1,960 | 73.0 |

The above are maximum efficiencies and cannot be achieved due to friction losses in the system. Significant implications of the above are:

Low temperature sources of energy such as solar energy cannot be converted efficiently unless they are concentrated to higher temperatures first.

Operating temperature limits of metals restrict the efficiency of internal combustion engines (much research is being done to develop engines with higher temperature tolerant components).

Turbines can operate at higher temperatures and are more efficient than internal combustion engines.

Examples of the limiting efficiency due to temperature and friction are the internal combustion engine (powers most of our automobiles, trucks and buses) which is on the order of 25 per cent for gasoline engines and 40 percent for diesel engines. The efficiency of a large combined cycle thermal electrical generating plant is on the order of 55 per cent of the heat input. The efficiency of airplane jet engines is on the order of 55 percent. In all cases a large portion of available energy is not realized.

The Second Law of Motion states the acceleration of a body is directly proportional to the force applied and inversely proportional to the mass of the body or:

$$A = F/M$$

or alternatively:

$$F = MA$$

Where: F equals force applied
M equals mass of the body (weight divided by
The gravitational constant)
A equals acceleration (rate of change in speed)

Energy is force applied to a moving object. From the above we can see that it takes more energy to accelerate a heavy body (start it into motion and increase speed). Also, it takes the same amount of energy to bring a body to rest as it does to put it in motion.

Consequences of the Second Law of Motion are most apparent in transportation;

For most automobiles, the weight of the car is much more than the weight of the occupants. Hence most of the energy is required by the weight of the automobile. From an energy perspective;

-Small, light automobiles are more energy efficient than heavy cars and adding more riders requires proportionately little added energy.

-Fully loaded buses, trains and planes are much more energy efficient than cars.

-Motorcycles are a very efficient form of individual transportation.

Much energy is lost braking a vehicle. Systems are being designed to capture and store this energy so it can be used to power the vehicle.

When thinking of uses and conversion of energy, it is important to keep the above principles in mind. Great strides are being made to approach the theoretical limits of efficiency and reduce the weight of transportation vehicles, but the limitations remain. As energy becomes more expensive, these principles will affect choices we make.

The first use of energy was probably early man's direct use of light and heat from naturally occurring forest fires. Eventually, people learned to start and control wood and biomass fires for warmth, cooking, and light. Today, energy is still used in these most basic of ways. In many areas of the world, they are the only forms of energy in use. In some areas, this is by choice due to economics, such as the use of wood-burning stoves in rural areas where wood is abundant. In others, this is for esthetics and pleasure, such as the use of a fireplace even though other forms of heat are available. Still, on balance, these uses account for less than 5 per cent of the world's current use of energy.

For most of the developed countries, energy-powered machines affect almost every facet of daily activity. Almost all machines are driven by conversion systems that convert primary sources of fuel to other forms of energy appropriate for particular uses. Following is a quick review of how these energy systems work.

## BOILERS

Early boilers were simply an open container with fire underneath to heat water. The first fuel was likely wood followed by coal. A major improvement was enclosing the container allowing steam to be generated. Further improvements were made by moving the fire inside a firebox where it came in contact with pipes containing water greatly increasing the efficiency of the system.

The utility of boilers was greatly expanded as suppliers of steam to turbines (will be discussed later) capable of driving generators and other machinery. Even today, boilers and steam play a major role in energy conversion and use. Current fuels for boilers include natural gas, coal, bunker fuel (lightly refined crude oil), and biomass such as wood-waste products. A major enhancement of boilers is to recover waste heat from other processes to improve efficiency of the total process.

## INTERNAL COMBUSTION ENGINE

Internal combustion engines convert heat generated by combustion of the fuel used (gasoline, diesel, etc.) to mechanical power.

All internal combustion engines operate on the same basic principle. A crankshaft drives a piston in a cylinder to compress, air and fuel to facilitate combustion. When combustion occurs pressure of resulting hot gases on the piston drives the crankshaft delivering power in the form of a rotating shaft.

Internal combustion engines remain the most widely used mechanisms for converting energy into a usable form. They are employed in automobiles, trains, airplanes, boats, ships, lawn mowers, power saws, tractors, and significant other applications. Normally, they are powered by gasoline or diesel fuel, but other fuels, including compressed natural gas or natural-gas liquids are also used.

## TURBINES

Probably the earliest form of energy conversion was some type of turbine. Turbines convert the energy of a moving stream of liquid (air, water, steam or gas) to rotate a shaft which is used to drive other machines. The first might have been a waterwheel where the water was captured on the circumference of the wheel, the weight causing the wheel to rotate.

In another form of a waterwheel, a stream of water is directed to strike the blades on the wheel. The pressure of the water striking the blades causes the wheel to rotate.

Vanes attached to a rotor (axle) parallel to the flow of a liquid or gas greatly improved turbine technology. Vanes completely encircle the rotor and can either take energy from a moving stream to rotate the shaft (engine) or add energy to a stream (compressor).

Turbines utilizing moving water to energize vanes attached to a rotor were an early application. They are still in wide use today and are central to hydroelectric plants driving generators for electricity.

Combustion gases or gases from external sources are widely used to drive vanes in numerous and varied applications.

A self-contained thermal turbine engine can be made by utilizing turbines on a single shaft for both compression of air and fuel for combustion and power delivery. The compressor end compresses incoming air to sufficient pressure for combustion of the air/fuel mixture. The hot expanding combustion gases drives the rotor end suppling rotating power to other machines. Exhaust out the back develops thrust.

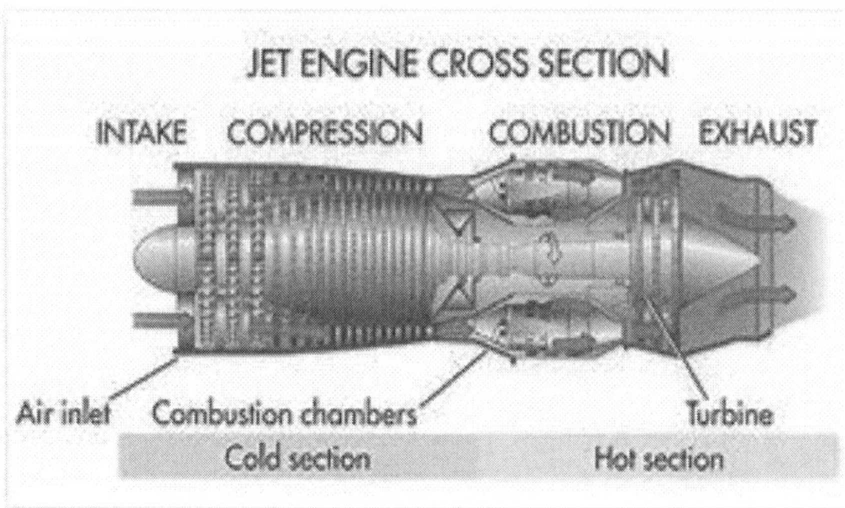


If the desire is to provide rotary power to a stationary source, turbine design maximizes capture of rotating energy and minimizes losses out the exhaust. If the desire is to power an airplane, thrust out the back is maximized (jet engine). Natural gas or liquid fuels such as aviation fuel and other light oils are used to fuel thermal turbines.

Turbine-based machines are simple in concept, have few moving parts, and are versatile for many applications. These traits make turbine-based machines the choice for many and varied applications.

When the application is to simply convert the kinetic energy of a moving stream of water, gas, or liquid to another form, turbines are very efficient—in the upper 90 per cent range. When the application is to convert thermal energy (from a burning fuel), the efficiency is about 45%. By adding a boiler to take waste heat from a turbine to generate steam (combined cycle) the efficiency can approach 55%.

## NUCLEAR REACTOR

Nuclear reactors generate heat from fission of nuclear fuel (fissile material). Fission occurs when a fissile atom captures a free neutron. When this happens the atom splits into two new atoms, releases neutrons, and produces heat. The free neutrons then react with other fissile atoms, continuing the process. Key control parameters become:

Control of free neutrons—too many free neutrons and the process goes too fast posing a safety hazard; too few and the process is slowed or stopped.

Removal of the heat generated in the reactor—to keep the reactor from overheating and at the same time, capturing heat for use.

To accomplish the above, a reactor, in its simplest form, is designed around four essential elements: fuel rods of fissile material; a moderator to even the movement of free neutrons; control rods to regulate the number of free neutrons; and a coolant to keep the reactor cool and take away the heat generated for capture and future use.

Primary control is managed by control rods which absorb neutrons. To slow the process down, they are inserted into the reactor to absorb free neutrons. To speed the process up they are pulled out to leave more free neutrons available for fission of the fuel.

Temperature is managed by adjusting the coolant circulation rate. The coolant captures heat from the process which is used to generate steam which drives turbines to generate electricity; hence temperature control also controls reactor output.

In some early designs, coolant fluids were also the power fluid. They were known as boiling water reactors. Later designs utilize a standalone cooling circuit and captured heat is transferred to the power fluid through heat exchangers.

Note in the above, there are no emissions from a nuclear reactor except heat. There are no air or water pollutants. This makes nuclear energy one of the most attractive energy forms available to us except for the following;

1. The spent fuel rods must be removed periodically. This waste contains a mixture of radioactive elements including plutonium which can be made into weapons of mass destruction. Management of this waste is important; however, the most problematic waste from a sizable reactor is less than one cubic yard per year.
2. Control of the process is critical. The consequence of the process going out of control is significant. Extreme attention

must be paid to design, construction, and operating parameters.

Heat removed from a nuclear reactor is used to make steam. This steam drives turbines to power applications.

In nuclear electricity plants the steam powers generators to produce electricity. Hence, the output that we use is electricity.

## GENERATORS AND MOTORS (DYNAMOS)

Perhaps the dynamo is the most versatile energy conversion machine. It converts mechanical energy into electrical energy and vice versa. When electricity is applied to produce mechanical power, the dynamo functions as a motor. When mechanical power is applied to produce electricity, the dynamo functions as a generator. The dynamo works as a result of two electrostatic forces:

- When electrical current is passed through a wire wound on an iron rod, a magnetic field is created.
- When a loop of conductive wire is rotated in a magnetic field, a current is created (generator). If an electric current is passed through the loop, the loop rotates (motor).

The magnetic field can be generated by a permanent magnet or by applying an electric current using the first principle. In most motors and generators, the magnetic field is formed by applying electricity creating a more powerful magnetic field and adding another level of control to the design. A dynamo is very efficient on the order of 95%.

A dynamo simply converts energy from one form to another utilizing electricity. Electricity is the most versatile form of energy, but it is not a primary source. A primary source of energy must be used to generate electrical energy. It is important to remember this when thinking of energy sources.

Five general systems or energy complexes are used to generate electricity: hydro, thermal, nuclear, wind, and solar. Since electricity is so important in today's energy picture, these systems will be discussed briefly.

## HYDROELECTRIC PLANTS

Hydroelectric plants convert stored energy of water behind a dam to electrical energy by utilizing a water turbine to drive an electrical generator as water is released from the dam. When elevation below the dam decreases the turbine and generation can be located lower by transporting the water through pipes. This increases the water pressure (head) and increases the amount of electricity that can be generated.

Almost all hydropower is converted to electricity today. Little is used directly to power machines.

## THERMAL ELECTICAL GENERATING PLANTS

Four types of thermal systems are used to generate electricity: Process gases powering turbines which drive generators; internal combustion engines driving generators; combustion gases powering integral turbine generator combinations, and boilers providing steam to power integral turbine generator combinations (combinations may be used is some systems to improve overall efficiency).

Internal combustion engines are fueled by gasoline, diesel or natural gas. Integral turbine generator combinations are fueled by natural gas or light refined crude. Boilers are fueled by natural gas, coal, lightly refined crude oil, or biomass.

## WIND

A wind turbine is used to extract energy from wind to drive a generator. The turbine looks like a huge airplane propeller. These are cropping up all over the world. No fuel is required, and electricity is the only output with no emissions.

## SOLAR

Electricity is generated from solar energy by two methods: direct conversions through photovoltaics or thermal conversion

As indicated, a photovoltaic plate can directly convert solar energy into electricity. The technology is proven, and great strides are being made in improving the efficiency of the conversion process and in reducing the cost. As these improvements continue and as the cost of energy rises, this source will become more competitive in price.

Thermal conversion of solar energy is difficult, since the temperature generated from absorbing sunlight is low; a person can stand in direct sunlight and not be harmed.

To achieve reasonable efficiencies the temperature of the system must be raised. This is done with large mirrors to concentrate the solar energy on boilers to generate steam. This steam is used to power a steam turbine which drives a generator; the only output is electricity with no emissions.

## FUEL CELLS

A fuel cell generates electricity by chemically combining hydrogen with oxygen. It consists of two electrodes (an anode and a cathode) that sandwich an electrolyte that allows ions to pass but blocks other electrons.

A fuel containing hydrogen is fed to the anode where the hydrogen electrons are freed leaving positively charged ions. The electrons travel through an external circuit as electricity while the ions diffuse through the electrolyte. At the cathode, the electrons combine with the hydrogen ions and oxygen to form water, the byproduct. Output from a fuel cell is electricity and water.

Fuel cells are a secondary source of energy as primary sources of fuel must be used to generate fuel for the cells. A convenient way to obtain the fuels is to apply electricity to water separating it into its components of hydrogen and oxygen. As electricity is the energy output of fuel cells this electricity can be used to generate hydrogen and oxygen. Some say free fuel can be obtained this way; however, we know that no more hydrogen and oxygen can be produced than consumed since to do so would result in a perpetual motion machine which is impossible. The end result is that no more energy can be produced than is used to make the fuel.

Nonetheless, fuel cells will play a significant role in the overall future energy picture as they are environmentally friendly, producing only water vapor as a waste product. Fuel cells can provide an alternative to fossil fuels for ground transportation.

Intermittent sources of energy such as wind, solar, or tidal can be used to generate hydrogen which can be stored for later use in fuel cells.

# SUMMARY OF ENERGY SOURCES AND USES

Prior sections dealt with sources of energy, how it is converted from one form to another and its uses. This section summarizes primary energy sources and uses.

Primary sources are grouped into six areas for convenience: Petroleum (crude oil); Natural gas; coal; nuclear; wind and other renewable. End uses are far too numerous to list but a fairly complete picture can be seen by breaking energy use into three categories; transportation, other and electrical generation.

Transportation required 24.3 percent of the world's primary sources of energy in 2018. Electricity generation required 50 percent of the world's primary sources of energy in 2018. These two uses required 74.3 percent of the world's energy in 2018. Due to their importance expanded data is presented for transportation and electricity generation in the second table.

Data were taken from several sources. There may be 1 to 2 percent variance in the data. Potential differences this small will not affect the significance of major conclusions.

SUMMARY OF ENERGY USES 2018

| | | Use of Primary Energy (Percent) | | |
|---|---|---|---|---|
| Primary Source | Percent of world energy Supply 2018 | Transportation | Other | Electricity Generation |
| Petroleum | 33.6 | 70 | 27 | 3 |
| Natural Gas | 23.9 | 2 | 58 | 40 |
| Coal | 27.2 | 1 | 11 | 88 |
| Nuclear | 4.4 | | | 100 |
| Renewables | | | | |
| Hydro | 6.4 | | | 100 |
| Wind | 2.5 | | | 100 |
| Other Renewable | 2 | | | 100 |

Note: consumption of bio diesel and ethanol included in petroleum data

SUMMARY OF TRANSPORTATION AND ELECTRICITY GENERATION 2018

| Primary Source | Percent of world energy Supply 2018 | TRANSPORTATION By Fuel % of World Energy | % by Source | ELECTRICITY GENERATION By Fuel % of World Energy | % by Source |
|---|---|---|---|---|---|
| Petroleum | 33.6 | 23.5 | 90+ | 1.0 | 3.0 |
| Natural Gas | 23.9 | 0.5 | | 9.6 | 23.2 |
| Coal | 27.2 | 0.3 | | 23.9 | 38.0 |
| Nuclear | 4.4 | | | 4.4 | 10.2 |
| Renewables | | | | 0.0 | |
| Hydro | 6.4 | | | 6.4 | 15.6 |
| Wind | 2.5 | | | 2.5 | 6.0 |
| Other Renewable | 2 | | | 2.0 | 4.0 |
| Totals | 100 | 24.3 | | 49.8 | 100.0 |

## MAJOR CONCLUSIONS

1. In 2018 85 percent of the world's energy was supplied by fossil fuels.
2. In 2018 90 plus percent of transportation fuels came from petroleum.
3. In 2018 74 percent of the worlds primary energy was used for transportation and electricity generation. One half for electricity generation and 24 percent for transportation.
4. In 2018 74 percent of electricity was generated with fossil fuels. Coal generated 38 percent.

## CONSUMPTION, PRODUCTION AND RESERVES

Energy must be available to consumers at reasonable prices to sustain and grow world economies. Economic activity in turn influences energy availability and price. Higher economic activity increases energy demand and price and lower economic activity reduces demand and lessens price pressure.

Price and availability are also related to technology, as some sources of energy are more expensive to develop and produce than others. Historically least expensive sources have been produced first. More expensive and technically complex sources were brought to market as prices reached levels making them economical to produce. Concurrently, technology advances reduced the cost of many sources of energy and allowed more complex sources to be produced with little or no increase in price.

Additionally, owners of energy must be willing to make it available to the marketplace at fair prices. Management of crude oil production and price by The Organization of Petroleum Exporting Countries (OPEC) is an excellent example of this. If the price of crude oil drops below a level they consider to be too low, availability to the market is reduced until an acceptable price is reached. Action by OPEC has tended to stabilize world crude oil prices during a time of excess crude oil production capacity. This allowed many sources of energy to be produced that might not have been economical if prices had been allowed to fall to free market levels during periods of excess capacity.

More recently, marketers in the supply chain between producers and consumers have become more of a factor in determining price and availability of energy. This leads to volatility in prices during periods of real or perceived shortages or oversupply. An example is extreme seasonal swings in natural gas prices in the United States.

The above complexities of economic activity, technology advances, and market forces make forecasts of total energy availability for the future difficult. However, plans must be made for future energy supplies to minimize impacts of disruptions.

The first step in developing a future plan is to look at today's situation. This can be simplified somewhat by looking at population, energy consumption, energy production and energy reserves by primary fuel source sorted into similar economic and/or energy regions.

2018 data is available for population, production and consumption of energy by source plus CO2 emissions. Data for charts that follow were taken from "BP Statistical Review of World Energy 2019".

Regions chosen are Canada; United States; Venezuela; Mexico, Central America and South America except Venezuela; Europe; Middle East; Commonwealth of Independent States (includes Russia); China; India; Asia Pacific less China and India and Africa.

To predict future energy supplies reserves of fossil fuels must be estimated as they are not replaced on an ongoing basis such as renewable sources.

Estimates of reserves vary by source and definition. Reserves are normally broken down into four categories; proven, probable, revisions to prior estimates and undiscovered. For the near-term it is prudent to use only proven estimates of reserves which are reflected in the following charts.

Reserves from probable, revisions to prior estimates and future discoveries are significant and may exceed today's estimates of proved reserves. However, timing, cost and environmental impact of development are uncertain.

Large reserves are in Canada's Alberta Tar Sands and Venezuela's Orinoco heavy oil deposits. This is why Canada and Venezuela are separated in the charts. Technology for both are proven but extraction will require large upfront capital and high operating costs.

A few comments on the United States are in order. The United States has increased crude production significantly in recent years utilizing horizontal wells and fracking technology. Since this is new technology, reserves shown for the United States are likely understated as the full potential may not be recognized.

Data on each major source of energy follows. Data is presented for each major country grouping as a percentage of world total. For example, chart 8 reflects data for petroleum. Africa, the first country shown, has 16.9% of the world's population, consumes 4% of the world's petroleum, produces 8.6% of the world's petroleum and has 7.2% of the world's petroleum reserves.

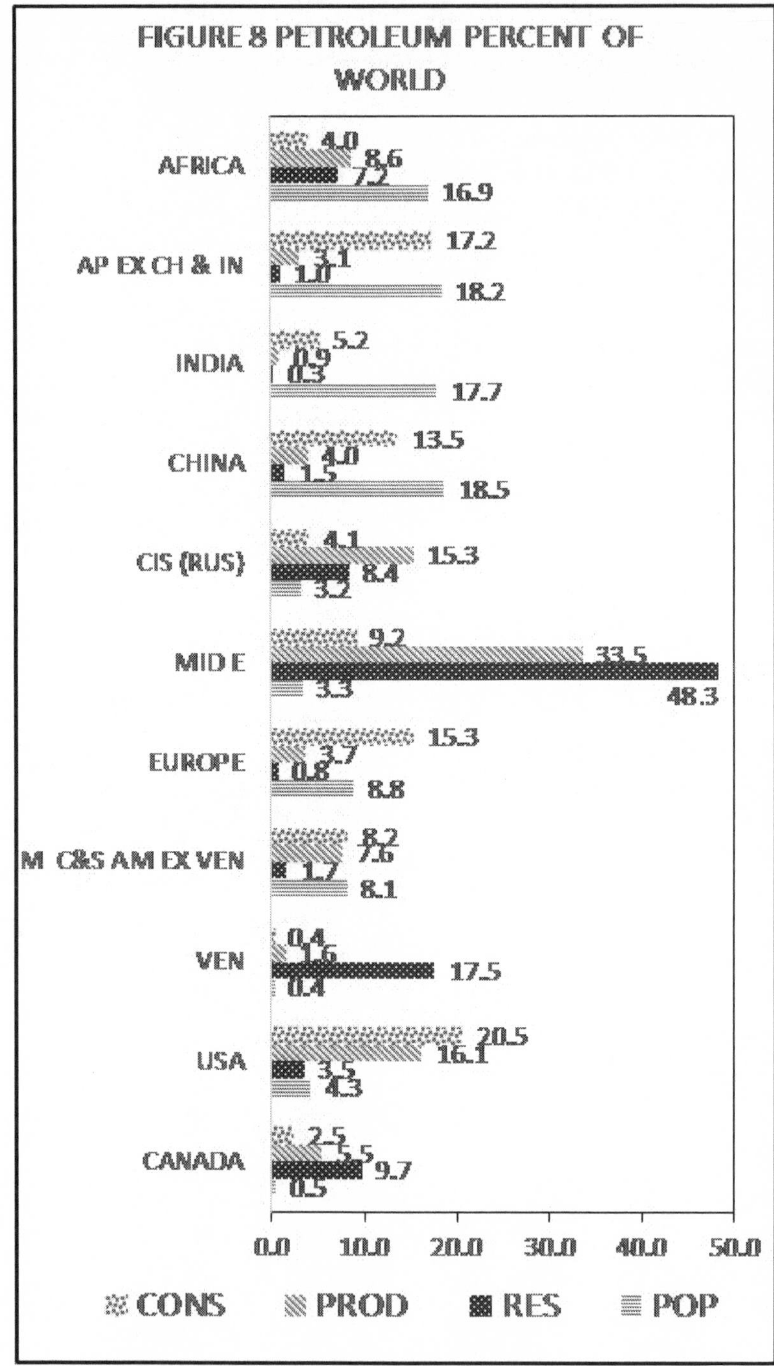
FIGURE 8 PETROLEUM PERCENT OF WORLD
AFRICA
4.0
8.6
7.2
16.9
AP EX CH & IN
17.2
3.1
1.0
18.2
INDIA
5.2
0.9
0.3
17.7
CHINA
13.5
4.0
1.5
18.5
CIS (RUS)
4.1
15.3
8.4
3.2
MID E
9.2
33.5
48.3
3.3
EUROPE
15.3
3.7
0.8
8.8
M C&S AM EX VEN
8.2
7.6
1.7
8.1
VEN
0.4
1.6
17.5
0.4
USA
20.5
16.1
3.5
4.3
CANADA
2.5
5.5
9.7
0.5
0.0
10.0
20.0
30.0
40.0
50.0
CONS
PROD
RES
POP

## PETROLEUM

Petroleum supplied almost 34% of the world's energy and 90 plus% of transportation fuels in 2018 making it the most important component in today's energy picture.

Figure 8 compares population to petroleum reserves, consumption, and production rates for eleven country groupings. The most striking thing demonstrated in figure 8 is the amount of reserves contained in the Middle East, Venezuela and Canada-75% with only 4.2 % of the world's population. The second most striking thing is that 63 percent of the world's population live in four major population areas where consumption exceeds production and there are few reserves available to improve the situation. The four areas are China, India, Europe and Asia Pacific.

Another important point in figure 8 is the low consumption of petroleum relative to population in developing countries. As they strive to improve their economies, additional pressure will be placed on world petroleum supplies.

The inescapable conclusion of the above is that net consuming countries will become more dependent on the Middle East for petroleum supplies so long as petroleum remains the fuel of choice.

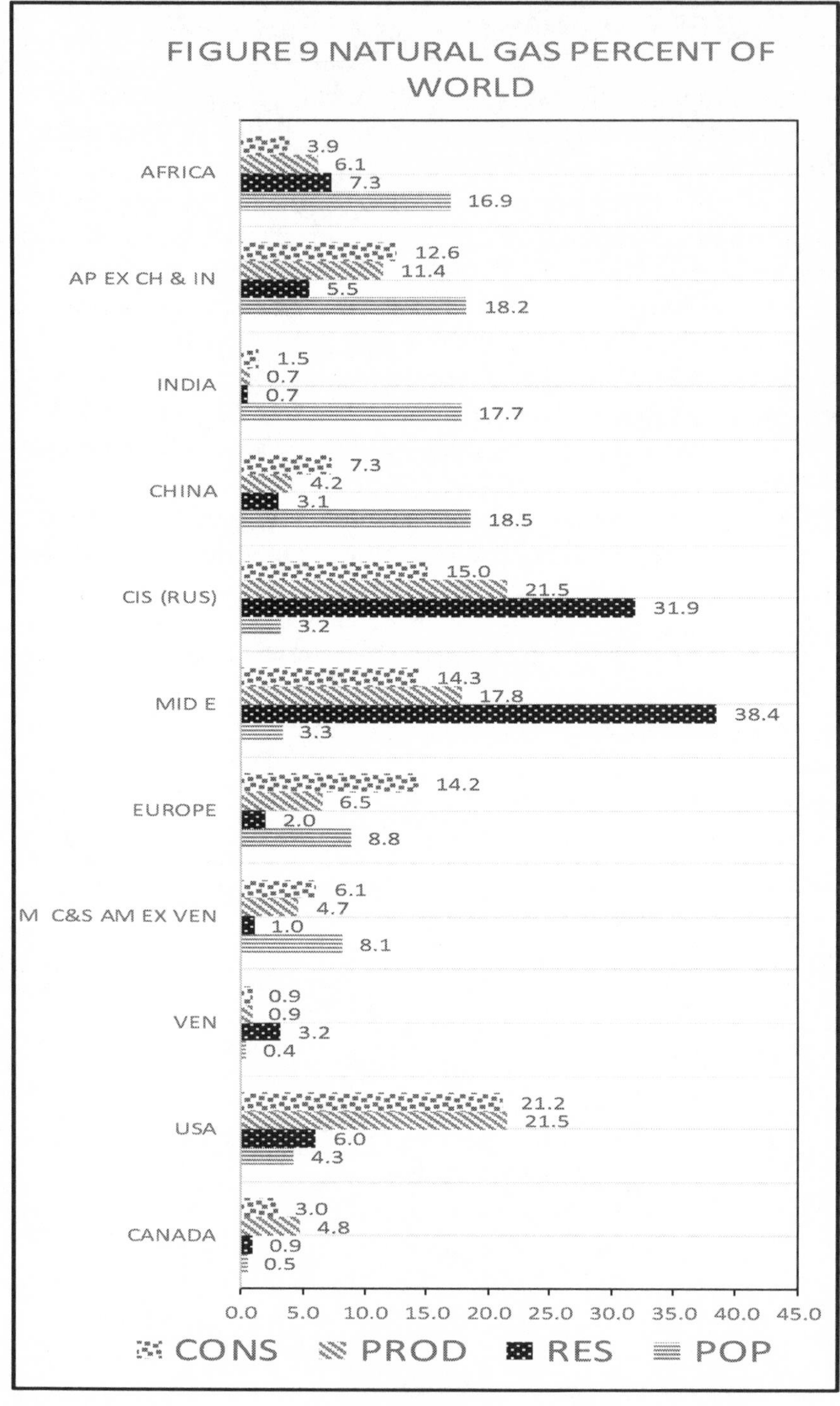

FIGURE 9 NATURAL GAS PERCENT OF WORLD
AFRICA
3.9
6.1
7.3
16.9
AP EX CH & IN
12.6
11.4
5.5
18.2
INDIA
1.5
0.7
0.7
17.7
CHINA
7.3
4.2
3.1
18.5
CIS (RUS)
15.0
21.5
31.9
3.2
MID E
14.3
17.8
38.4
3.3
EUROPE
14.2
6.5
2.0
8.8
M C&S AM EX VEN
6.1
4.7
1.0
8.1
VEN
0.9
0.9
3.2
0.4
USA
21.2
21.5
6.0
4.3
CANADA
3.0
4.8
0.9
0.5
0.0 5.0 10.0 15.0 20.0 25.0 30.0 35.0 40.0 45.0
CONS
PROD
RES
POP

## NATURAL GAS

Natural gas is the cleanest burning of the fossil fuels and currently supplies 24 percent of the world's energy.

Figure 9 compares population, reserves, consumption, and production rates for the same eleven country groupings.

The Middle East and The Commonwealth of Independent States (includes Russia) contains 70 percent of natural gas reserves with only 6.5 percent of the world's population.

The same four regions containing 63% of the world's population that are deficient comparing production to consumption for petroleum are deficient for natural gas also.

Countries that are net consumers of natural gas will become more dependent on the Middle East and CIS in the future.

Also, like petroleum, natural gas reserves for the United States is likely understated as the full potential of horizontal wells and fracking technology may not be recognized.

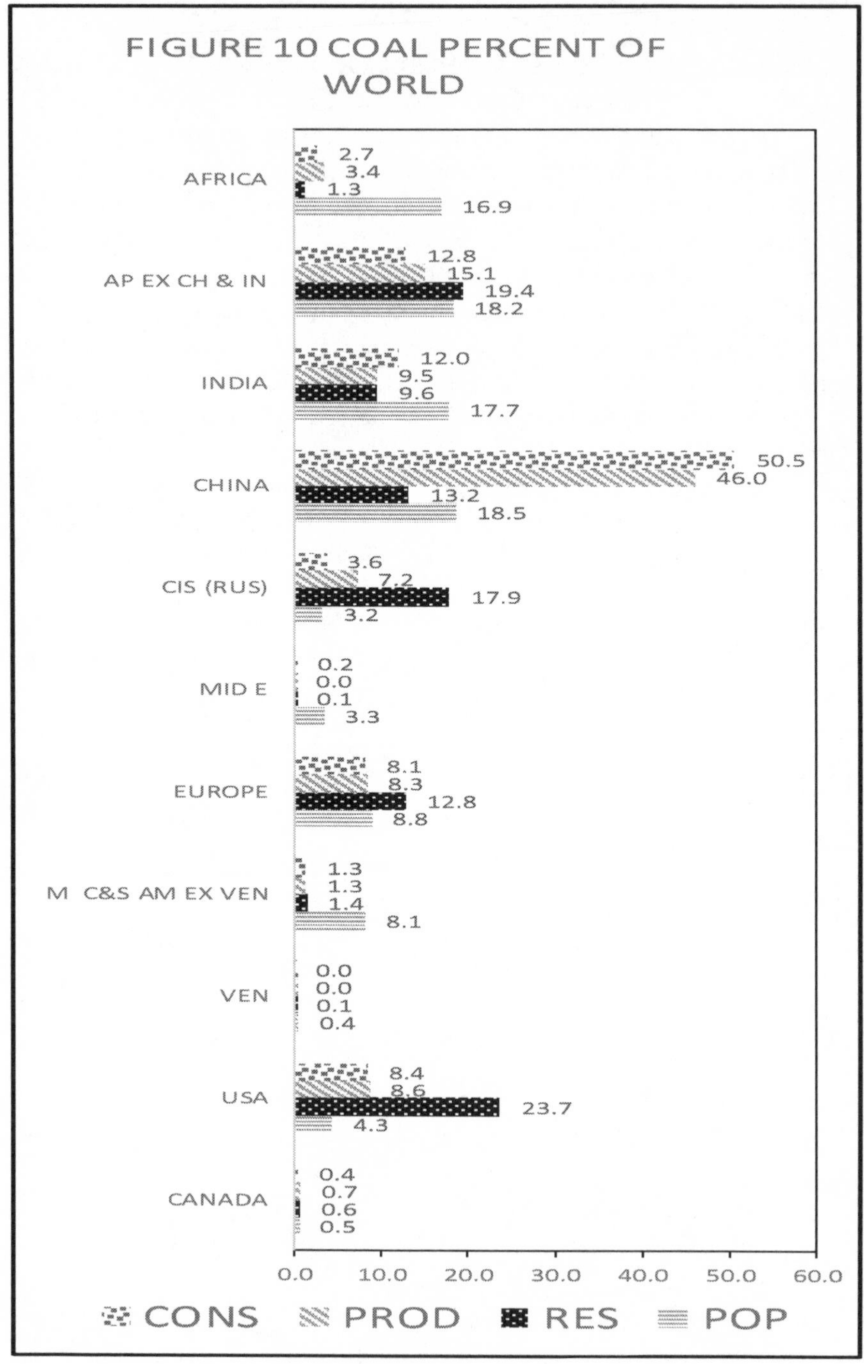
FIGURE 10 COAL PERCENT OF WORLD
AFRICA
2.7
3.4
1.3
16.9
AP EX CH & IN
12.8
15.1
19.4
18.2
INDIA
12.0
9.5
9.6
17.7
CHINA
50.5
46.0
13.2
18.5
CIS (RUS)
3.6
7.2
17.9
3.2
MID E
0.2
0.0
0.1
3.3
EUROPE
8.1
8.3
12.8
8.8
M C&S AM EX VEN
1.3
1.3
1.4
8.1
VEN
0.0
0.0
0.1
0.4
USA
8.4
8.6
23.7
4.3
CANADA
0.4
0.7
0.6
0.5
0.0
10.0
20.0
30.0
40.0
50.0
60.0
CONS
PROD
RES
POP

## COAL

Coal has been the world's largest source of energy for the last one hundred thirty years as shown in figure 5 and currently supplies 27%. Figure 10 reflects the current distribution of reserves, production and consumption compared to population for the eleven country groupings.

China, India and Asia Pacific less China and India consume 75 percent of the world's coal with 54 percent of the world's population.

Coal currently supplies 58 percent of China's energy, but they only have 13 percent of the world's coal reserves. This will severely impact China's energy picture as their coal reserves are depleted.

The United States contains 23.7 percent of the world's coal reserves. With clean burn technology and CO2 recovery this could be a significant plus for the United States.

In other areas production and consumption are for the most part in balance.

Reliance on coal as their major source of energy in China, India and Asia Pacific is unlikely to change significantly in the near term as they have few petroleum and natural gas reserves.

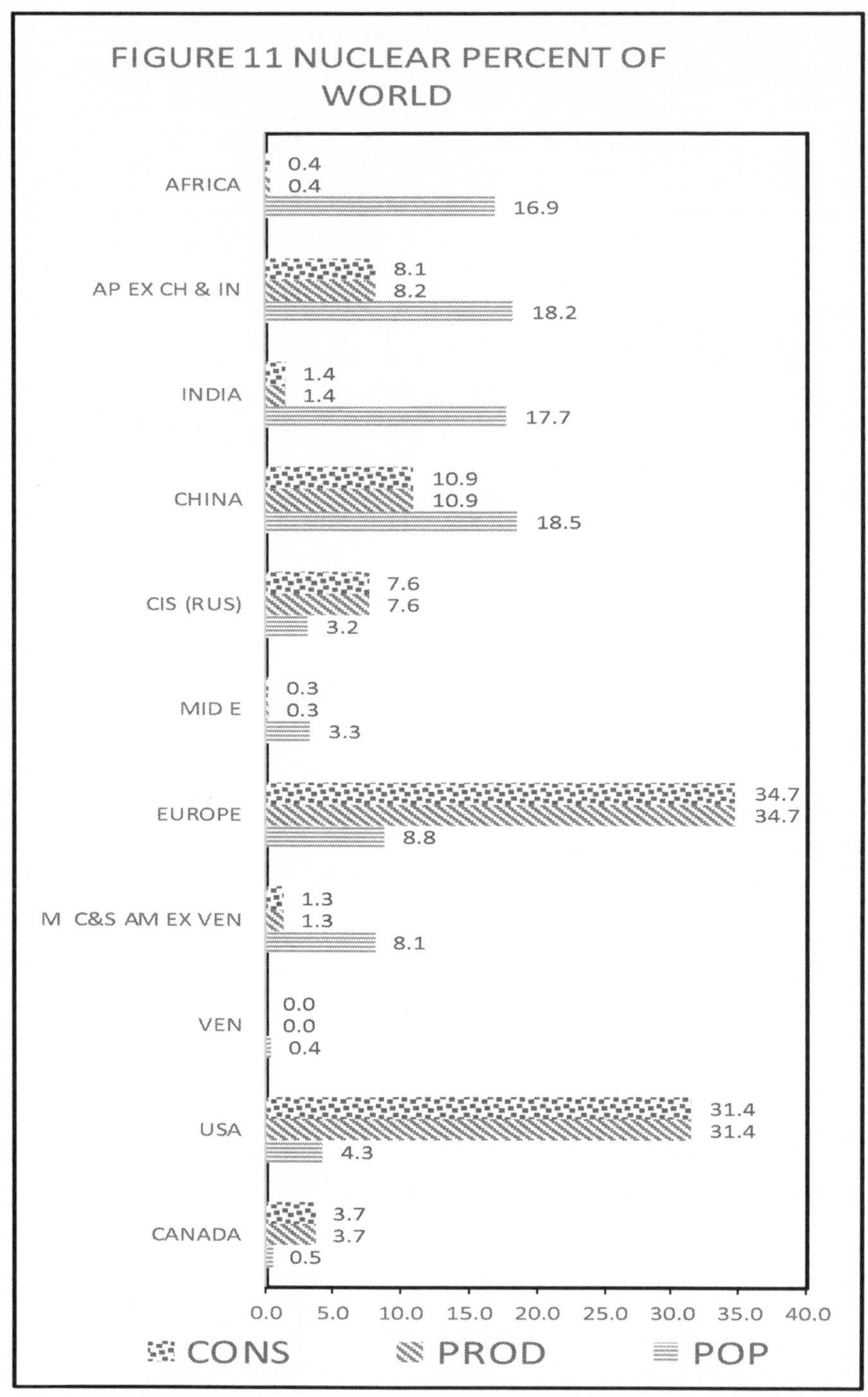
FIGURE 11 NUCLEAR PERCENT OF WORLD
AFRICA
0.4
0.4
16.9
AP EX CH & IN
8.1
8.2
18.2
INDIA
1.4
1.4
17.7
CHINA
10.9
10.9
18.5
CIS (RUS)
7.6
7.6
3.2
MID E
0.3
0.3
3.3
EUROPE
34.7
34.7
8.8
M C&S AM EX VEN
1.3
1.3
8.1
VEN
0.0
0.0
0.4
USA
31.4
31.4
4.3
CANADA
3.7
3.7
0.5
0.0 5.0 10.0 15.0 20.0 25.0 30.0 35.0 40.0
CONS
PROD
POP

## NUCLEAR

Statistics for nuclear which supplied 4.4% of the world's energy in 2018 are shown in Figure 11.

. As shown, Europe and the United States produce significant amounts of nuclear energy relative to their population.

No reserves are shown for nuclear because future amounts of energy derived from nuclear is dependent on technology used and safety/political issues which are difficult to predict.

Nuclear can have a significant role in future energy supplies if these technologies can be developed.

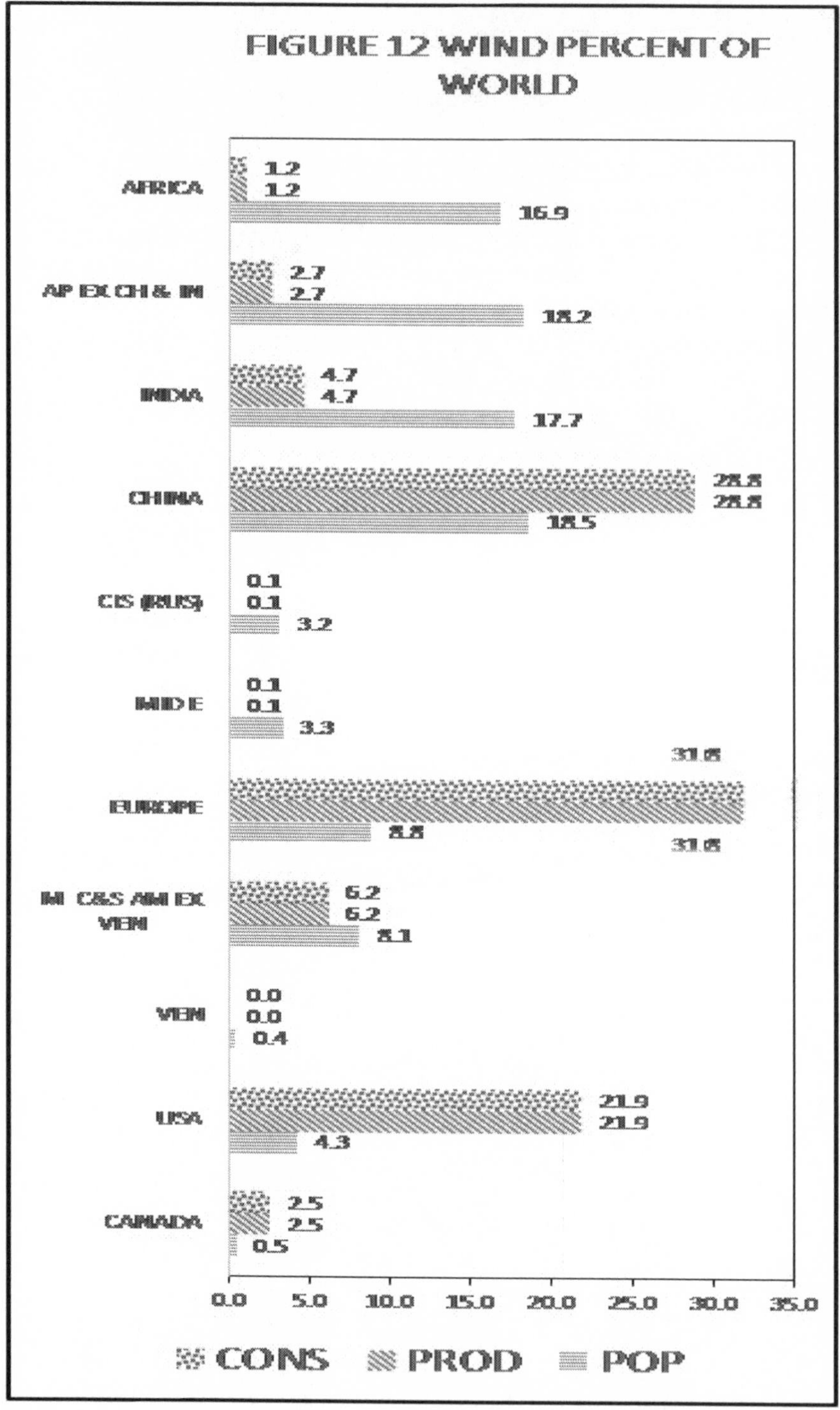
FIGURE 12 WIND PERCENT OF WORLD
AFRICA
1.2
1.2
16.9
AP EX CH & IN
2.7
2.7
18.2
INDIA
4.7
4.7
17.7
CHINA
28.8
28.8
18.5
CIS (RUS)
0.1
0.1
3.2
MID E
0.1
0.1
3.3
EUROPE
31.8
8.8
31.8
M C&S AM EX VEN
6.2
6.2
8.1
VEN
0.0
0.0
0.4
USA
21.9
21.9
4.3
CANADA
2.5
2.5
0.5
0.0
5.0
10.0
15.0
20.0
25.0
30.0
35.0
CONS
PROD
POP

## WIND

Statistics for wind, which supplied 2.5% of the world's energy in 2018, are shown in Figure 12.

Significant amounts of wind energy are generated in Europe, China and The United States.

Little is generated in the remainder of the world. Wind energy technology is simple and represents an excellent source of energy in developing countries.

No reserves are shown for wind energy but a simple ratio of population in China, Europe and the United Sates to total world population suggest wind has the potential to supply 15% of total world energy.

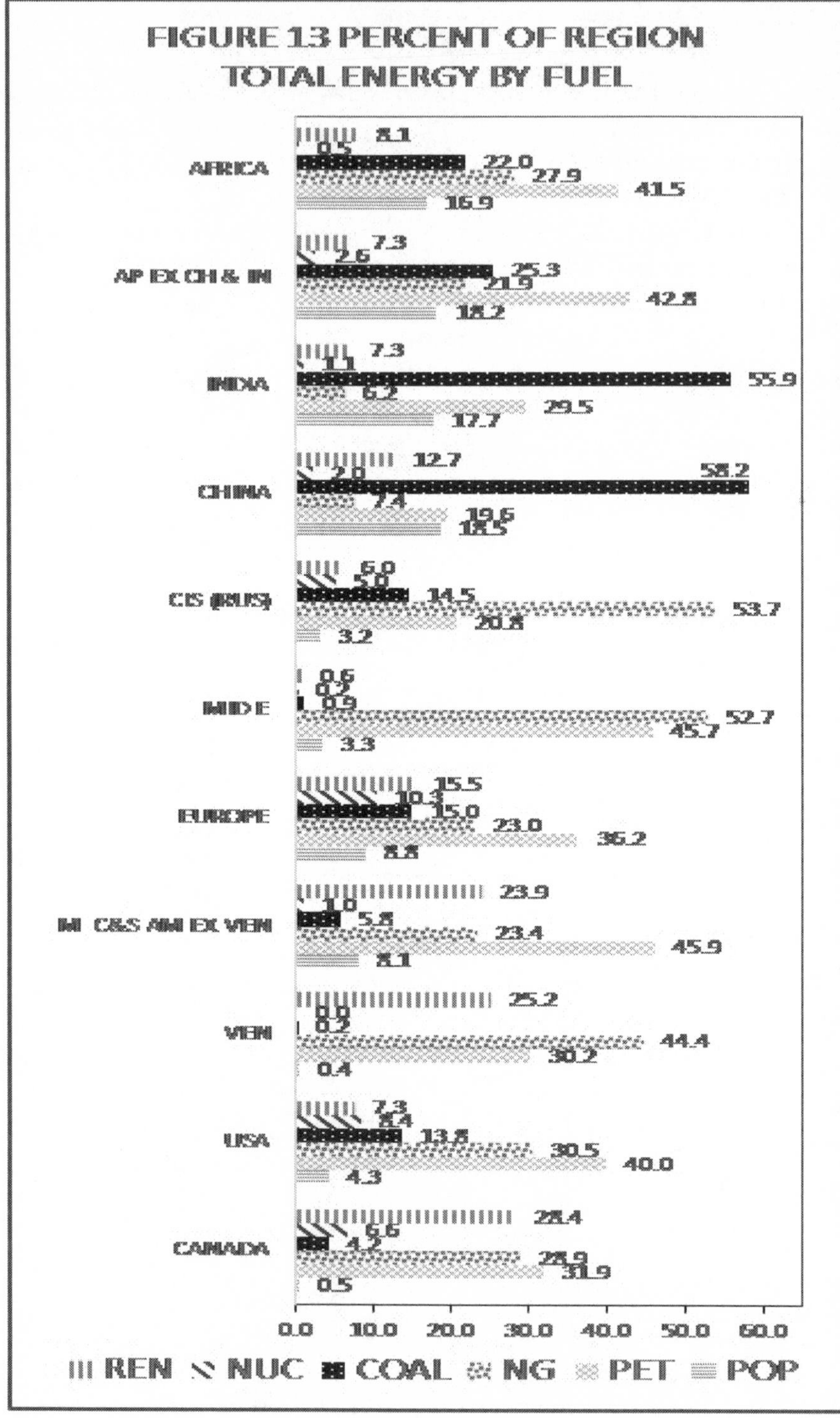
FIGURE 13 PERCENT OF REGION TOTAL ENERGY BY FUEL
AFRICA
AP EX CH & IN
INDIA
CHINA
CIS (RUS)
MID E
EUROPE
M C&S AM EX VEN
VEN
USA
CANADA
8.1
0.5
22.0
27.9
41.5
16.9
7.3
2.6
25.3
21.9
42.8
18.2
7.3
1.1
55.9
6.2
29.5
17.7
12.7
2.0
58.2
7.4
19.6
18.5
6.0
5.0
14.5
53.7
20.8
3.2
0.6
0.2
0.9
52.7
45.7
3.3
15.5
10.3
15.0
23.0
36.2
8.8
23.9
1.0
5.8
23.4
45.9
8.1
25.2
0.0
0.2
44.4
30.2
0.4
7.3
8.4
13.8
30.5
40.0
4.3
28.4
6.6
4.2
28.9
31.9
0.5
0.0
10.0
20.0
30.0
40.0
50.0
60.0
REN
NUC
COAL
NG
PET
POP

## REGION TOTAL ENERGY BY FUEL

Energy consumption by fuel for each of the eleven country groupings are shown in figure 13.

China and India, the two largest population centers, rely on coal for over 50% of their energy.

CIS countries including Russia, The Middle East and Venezuela are the largest consumers of natural gas as a share of their total energy needs. Natural gas is the cleanest burning fossil fuel.

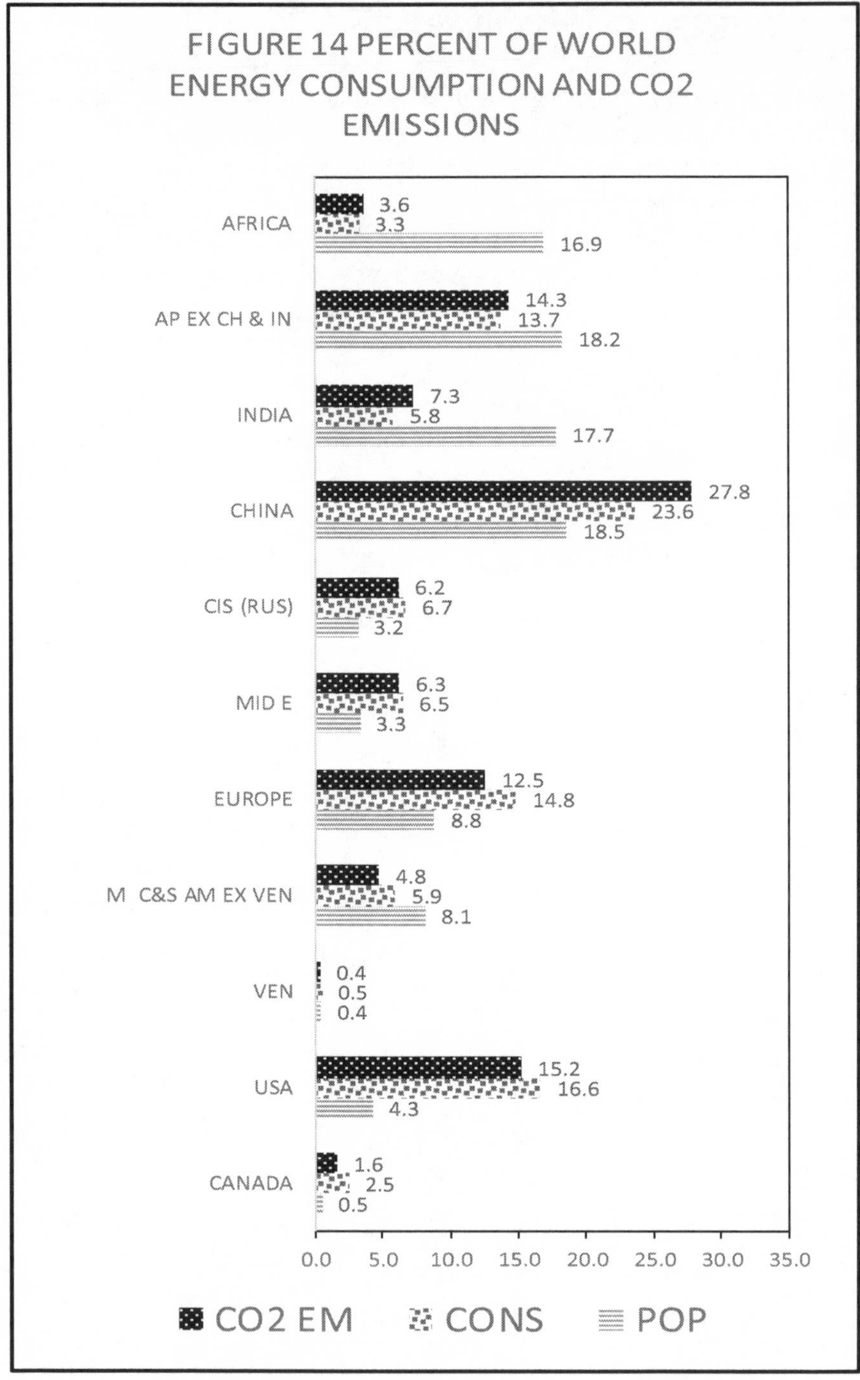
FIGURE 14 PERCENT OF WORLD ENERGY CONSUMPTION AND CO2 EMISSIONS
AFRICA
3.6
3.3
16.9
AP EX CH & IN
14.3
13.7
18.2
INDIA
7.3
5.8
17.7
CHINA
27.8
23.6
18.5
CIS (RUS)
6.2
6.7
3.2
MID E
6.3
6.5
3.3
EUROPE
12.5
14.8
8.8
M C&S AM EX VEN
4.8
5.9
8.1
VEN
0.4
0.5
0.4
USA
15.2
16.6
4.3
CANADA
1.6
2.5
0.5
0.0
5.0
10.0
15.0
20.0
25.0
30.0
35.0
CO2 EM
CONS
POP

## WORLD ENERGY CONSUMPTION AND CO2 EMISSIONS

Total energy consumption and resulting CO2 emission for each region are shown in Figure 14 as a percent of world total.

China is the largest consumer of energy and releases the most CO2 emissions into the atmosphere.

The United States is second. This appears out of order due to its relatively small population compared to the total world. However, the United states generates the highest GDP per person in the world and is the world's economic engine. If the United States has the sniffles, world economies catch a cold.

Data in the above section relate energy consumption by major fuel and resulting CO2 sources for eleven economic regions. Many and varied conclusions can be reached, too many to discuss, as their importance is related to the topic being considered. Rather this section should be a source of information to be analyzed as energy options are considered.

## A LOOK TO THE FUTURE

Potential world social, economic and political events as well as uncertainties in energy resource estimates make projections of future energy demand and supply difficult. However, to properly plan, estimates must be made and adjusted as required.

Figure 15 is world reference case population projections through 2050 from "United Nations World Population Prospects 2019".

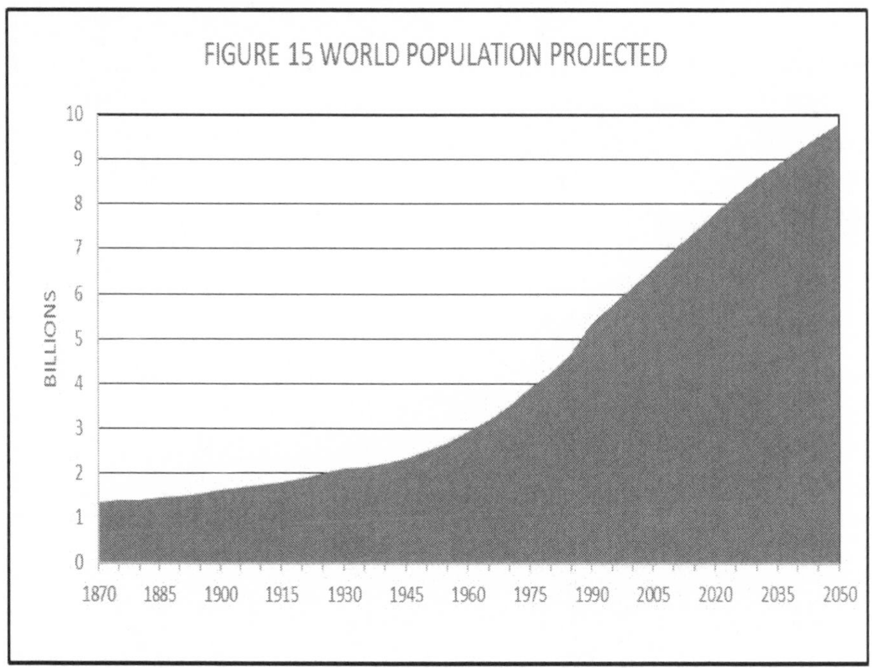


Growth in world population continues reaching 9.8 billion in 2050. There is only a slight decrease in the rate of increase toward the end of the period. This increasing population will need energy for essentials of life and jobs for their economic wellbeing.

Population is the basis for projecting world energy demand. The second step is melding population and economic forecasts to obtain energy demand.

The U.S. Energy Information Administration (eia) made world forecasts for energy consumption through 2050' in "International Energy Outlook 2017".

Several cases were developed. The reference case is the most likely given normal world expectations and was used in the discussion below.

Figures 16 and 17 reflect world energy consumption projected through 2050 for the reference case.

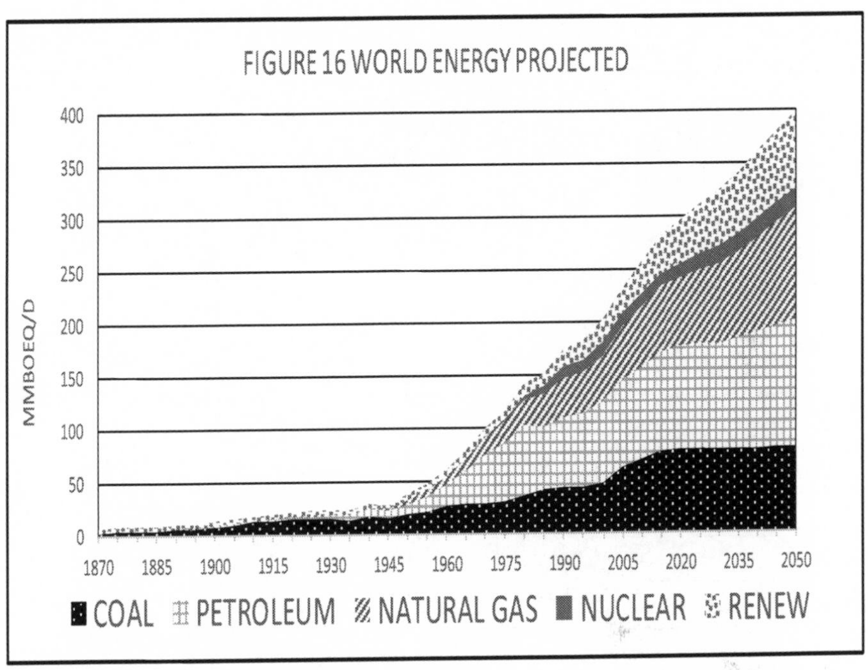

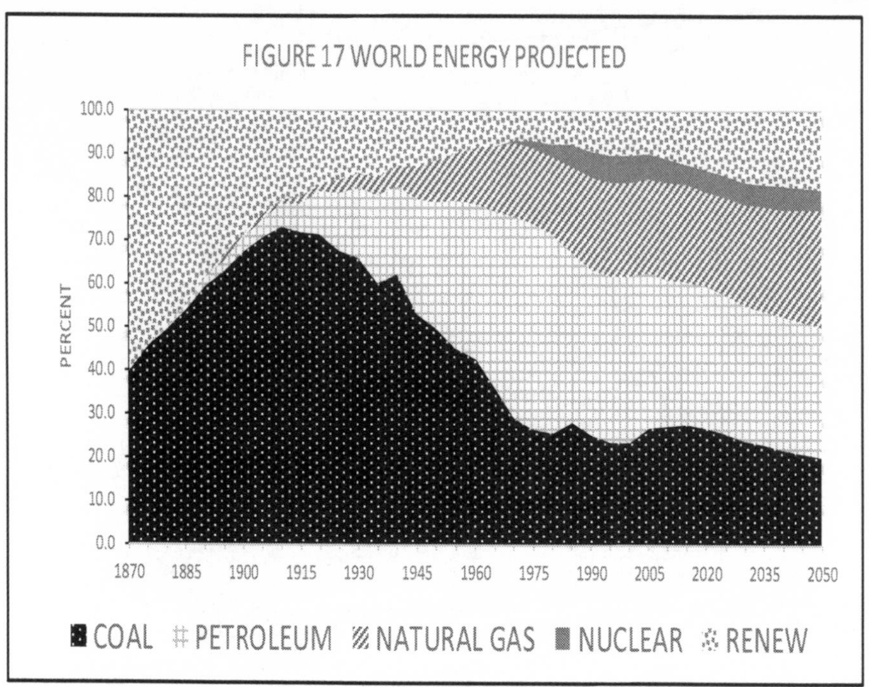


As shown all sources of energy consumption increase except coal which remains essentially flat.

The largest increase is renewable and as a result the percentage of energy supplied by renewables increases from 12.5% in 2015 to 18.0% in 2050. This is a significant increase but does not materially change the energy picture over the projected period. Fossil fuels supplied 83% in 2015 and is forecast to supply 77.1% in 2050.

As expected, $CO_2$ emissions into the atmosphere are forecast to rise as shown in Figure 18.

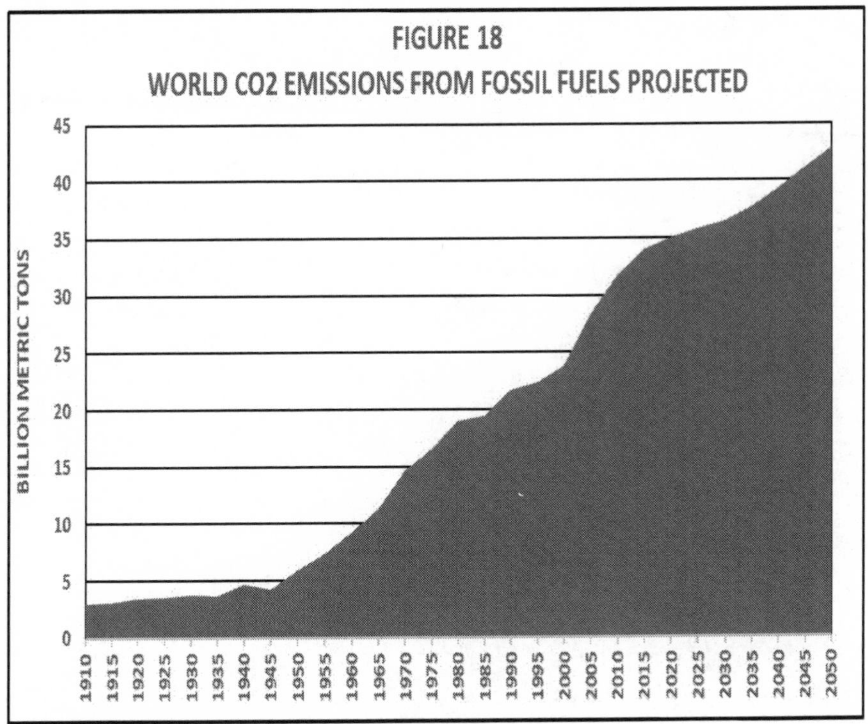


The above charts illustrate expected demand through 2050. The next step is to confirm energy supply can meet forecast demand. Petroleum and natural gas supplied 55% of the world's energy and over 90 % of transportation fuels in 2015 and have the lowest estimates of reserves relative to consumption. Due to their importance they are examined in more detail.

Petroleum and natural gas production in the near term will come from proven reserves. There are four categories of reserve estimates; proven, probable, revisions to prior estimates and undiscovered.

"Proven reserve" estimates are based on known accumulations using existing technology.

"Probable reserves" are based on known accumulations, but technical uncertainties prevent them from being classified as proved.

"Revisions to prior estimates" are based on changes in known accumulations due to performance. They can be positive or negative but historically they have been significantly positive. This is due to regulatory and financial reporting requirements that may restrict what can be included in proven and probable estimates. As technology improves performance can significantly increase estimates of reserves. Horizontal wells that are fracture treated are good examples of this. Since it is relative new technology the full potential of fractured treated horizontal wells is not necessarily included in proved and probable estimates.

"Undiscovered estimates" are risk discounted estimates of future discoveries. Significant upside exists but timing of discoveries, development and production are uncertain.

Accordingly, the prudent way to plan for future energy supplies is to base estimates on proven reserves in a stable world political/economic environment. The base plan can be adjusted as dictated by future world events and/or reserve estimates.

Production forecasts for petroleum and natural gas based on proved reserves are presented below due to their importance in the energy supply picture.

## PETROLEUM

Petroleum proved reserves are estimated at 1.729 trillion barrels in BP's "Statistical Review of World Energy June 2019". BP estimates were put together from many sources as discussed in their report. Further, proved reserves are not necessarily consistent with company proved estimates due to regulatory restrictions. Nonetheless their estimates represent a starting point for planning.

Two things are important in future forecasts; 1. the production profile and 2. time of peak production. A bell curve profile over the total production cycle (many things in nature take this form) was used to estimate remaining production. Methodology was to use the early part of the eia consumption profile until ½ ultimate recovery was produced. Peak production occurs at this point. The decline portion of the production profile roughly mirror the first part. A check for accuracy was made by comparing the resulting remaining production to estimated reserves. Results are shown in figure 19 below.

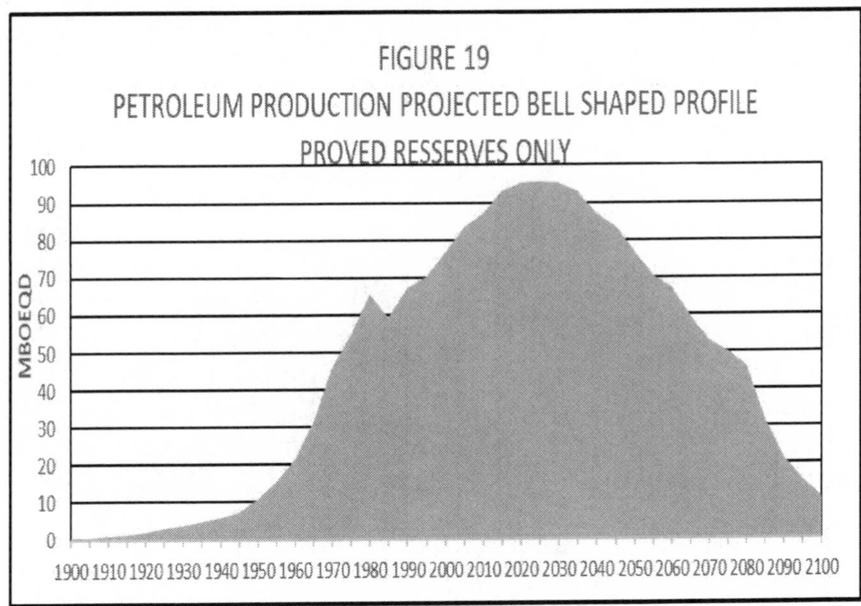


Peak production occurs at about 2030. The forecast does not meet the eia consumption forecast for petroleum toward the end of the period. This does not mean the forecast will not be met; just that some of the supply have come from probable, revisions to prior estimates, and undiscovered sources.

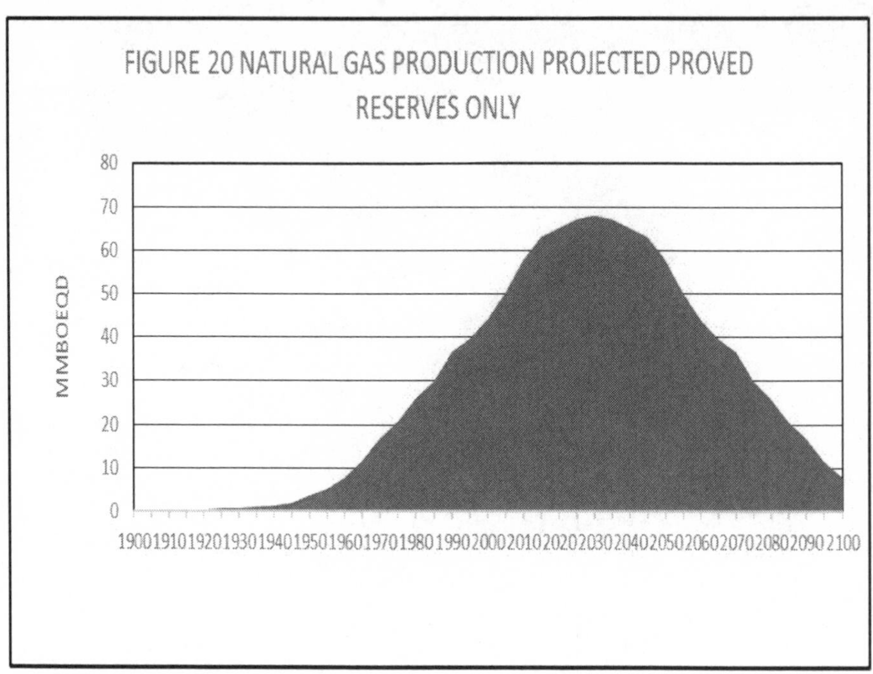


NATURAL GAS

World natural gas proved reserves are estimated at 6952 trillion cubic feet in BP's Statistical Review of World Energy 2019. The bell-shaped profile of total life production was used to estimate future production. Results are shown in figure 20. Production peaks past 2030. Like petroleum, natural gas production from proven reserves does not meet demand in the eia forecast through 2050. Part of the forecast will have to be met by production from probable, plus revisions to prior estimates and undiscovered reserves.

.

## PETROLEUM PLUS NATURAL GAS

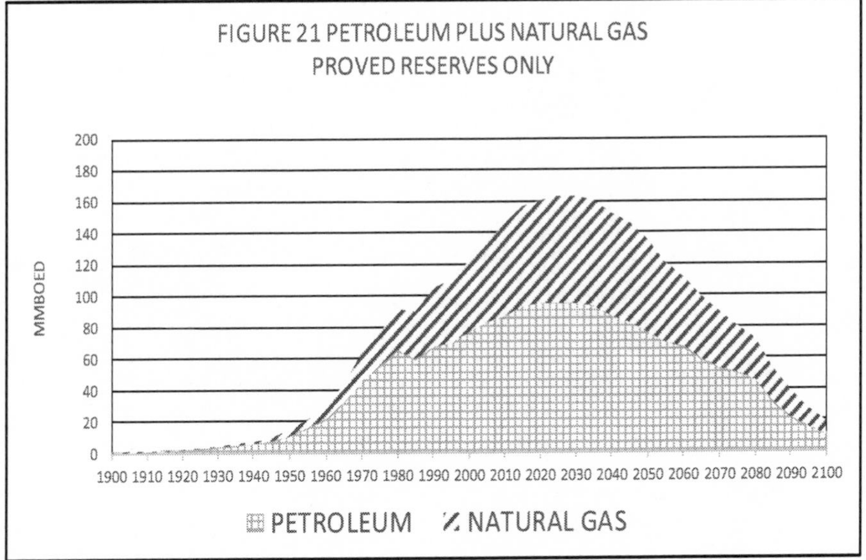


Figure 21 show petroleum and natural gas production added together

Forecast production from petroleum and natural gas proved reserves peak in 2030 at 163 million barrels equivalent per day (MMBOEQD); decline to 135 MMBOED in 2050 and 20 MMBOEQ in 2100.

Production from proven reserves does not meet the eia forecast in 2050 only 30 years in the future. Further, production from proven reserves declines by 87 % by 2100 which is only 80 years in the future.

ENERGY SOURCES PROJECTED TO 2100

Figure 21 illustrates the significance of petroleum and natural gas declining production from proven reserves by forecasting the total energy picture through 2100 with the following assumptions:

1. Total energy consumption is fixed at the eia forecast 2050 level.
2. Renewable, nuclear and coal sources are fixed at eia 2050 levels.
3. Petroleum and natural gas production shown is from proven reserves only.

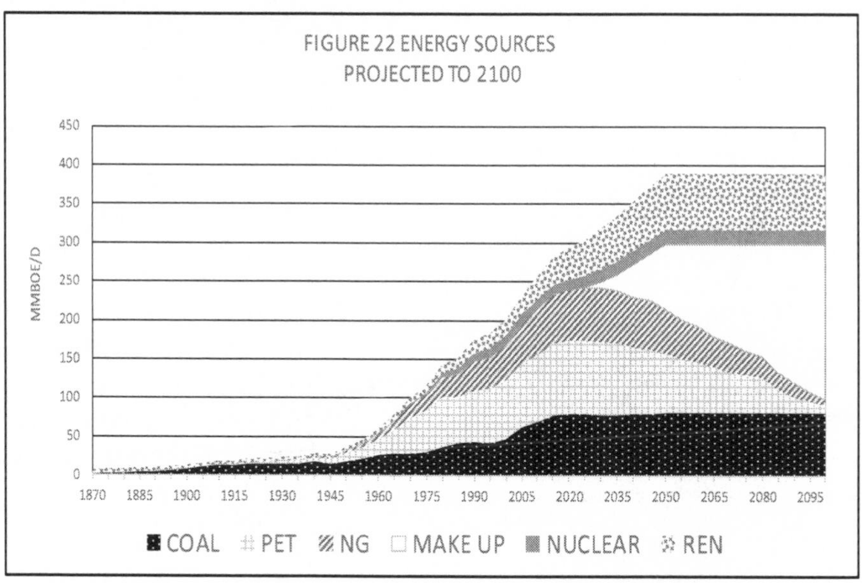


The area of the above chart labeled MAKE UP is significant. It represents the amount of energy that will have to be made up due to decline of production from proven petroleum and natural gas reserves. It will be necessary to develop 92 MMBOEQ/D by 2050 and 208 MMBOEQ/D by 2100 from unproven petroleum and natural gas reserves, coal, nuclear and/or renewable just to meet the eia 2050 forecasts. If these supplies are not developed, world economies will likely suffer.

The dilemma of future energy is illustrated above:

On one hand, if the desire is to reduce reliance on fossil fuels (due to climate change concerns) and petroleum and natural gas are reduced in a manner similar to the forecasts above, massive sources from nuclear and renewables must be developed. This will be a major undertaking and world economies will likely suffer. If climate change effects are not severe, huge capital and economic costs will have been incurred uselessly.

On the other hand, if petroleum and natural gas supplies from unproven sources are counted on but not realized in expected quantities, it will be necessary to develop nuclear and renewable sources on tight timetables. Shortages of energy will occur, harming world economies.

The only reasonable approach to the above dilemma is a reasoned-planned approach allowing for adjustments as indicated.

Supplying forecast future growing energy demand is truly a momentous undertaking. However, it is made somewhat easier because nuclear and most types of renewable (hydroelectric, wind and photovoltaics primarily) are captured as electricity initially. Since thermal energy from fossil fuels is converted to electricity at about 50% efficiency, it will only take one half the amount of these sources to offset one unit of petroleum and natural gas energy.

## CO2 EMISSIONS

CO2 emissions to the atmosphere from fossil fuels included in the eia forecast are shown in figure 18. Estimated emissions increase from 35 billion metric tons in 2015 to 43 billion metric tons in 2050.

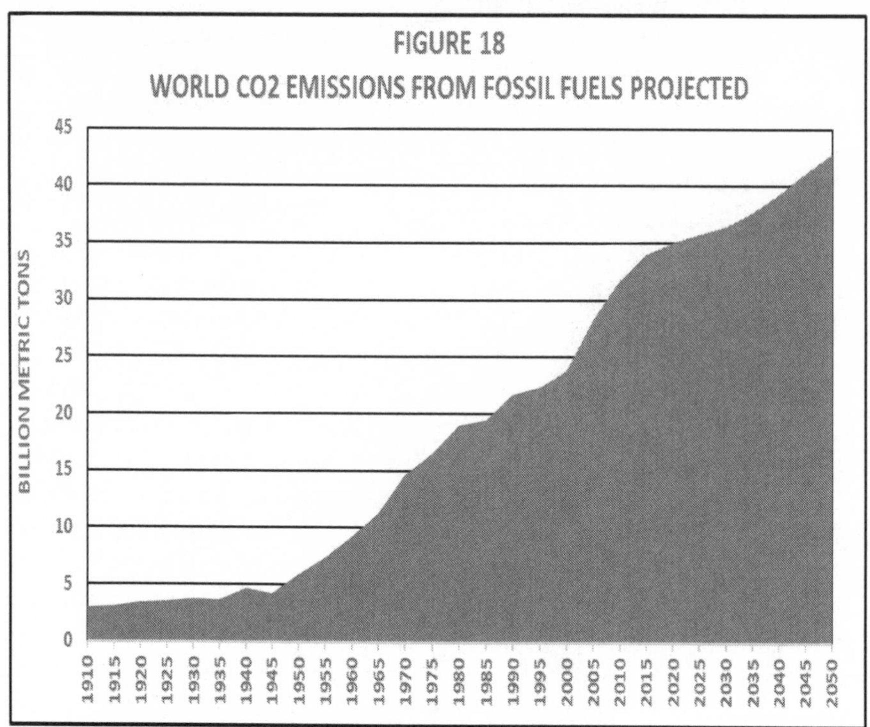


No progress is made reducing $CO_2$ emissions through the eia forecast period due to continued reliance on fossil fuels.

Nature removes $CO_2$ from the air through photosynthesis in plants. Since the concentration of $CO_2$ in the atmosphere is increasing, more $CO_2$ is being put into the atmosphere than plants alone can currently remove. Technologies are available that can assist nature in this effort.

## CO2 CAPTURE AND MITIGATION

Proven technology is currently in use to remove $CO_2$ from flue gases at industrial and electricity generating plants. Currently this captured $CO_2$ is used in industrial and food service applications or to recover crude oil left behind after normal recovery processes are over.

If negative effects of $CO_2$ build up in the atmosphere are shown to be significant enough this technology can be expanded, and the increased $CO_2$ recovered used in oil recovery processes or sequestered in underground reservoirs such as depleted oil and gas fields.

CO2 is miscible with some crude oils and formation water at reservoir temperature and pressure. This means the three liquids are mixable in all proportions. CO2 pumped into these reservoirs will mix with crude oil and water leaving CO2 behind. If this process is carried on long enough the oil and formation water will be displaced leaving the reservoir filled with CO2. On average, 5 barrels of CO2 will be left behind for each barrel of oil produced. The result is increased crude oil for transportation with a very positive CO2 balance.

Recovery of crude oil by CO2 injection has significant potential in the United States due to large coal reserves in the United States. Using clean burn technology and CO2 recovery, electrical energy can be obtained from coal fired generating plants. The captured CO2 can be used to recover petroleum for transportation fuels leaving CO2 sequestered in the reservoir.

Carbon Engineering, a Canadian company, has a pilot plant in operation that recovers CO2 directly from air. They are developing technology to make diesel fuels using the recovered CO2 and hydrogen recovered by electrolysis of water. If electricity for electrolysis is from renewable sources, then it is possible for the total cycle to near neutral relative to CO2.

Carbon Engineering are demonstrating innovative methods to have climate friendly energy sources.

Reforestation can greatly aid nature in removing CO2 from the atmosphere. Trees can be replanted where logs were harvested and not replaced. Many areas which were cleared for agriculture and are no longer in use can be replanted in trees. Areas that are too dry for vegetation today may benefit from increased rain and snow fall as warmer air from global warming absorbs more moisture from oceans and transport it onshore. These areas can be planted in trees. Planned replanting can decrease the time of regrowth as compared to nature alone.

Many options are available to reduce CO2 in the atmosphere if effects of global warming can be shown to justify the costs. Costs are not just monetary. Caution must be taken before wholesale changes are made as major economic and social disruptions could result.

## SUMMARY OF FOSSIL FUELS AND CO2 EMISSIONS PROJECTIONS

The above discussion suggests the following issues with associated potential impacts in the near future;

1. Heavy reliance on petroleum and natural gas. The longer the world depends on petroleum and natural gas in increasing amounts, the harder it will be to convert to other energy sources when necessary. World peak production capacity for each will likely occur in the next 25 years. If replacements for petroleum and natural gas are not in place beforehand, severe disruptions to world economies could result.
2. Increasing concentration of petroleum and natural gas supplies in regions with large reserves. The Middle East, CIS (Russia), Canada and Venezuela contain reserves of petroleum and natural gas far in disproportion to their population and consumption. Tightening petroleum and natural gas supplies as world demand approaches world production capacity will only make this issue more critical.
3. Increasing CO2 emissions and CO2 concentration build up in the atmosphere. Higher levels of CO2 in the atmosphere make continued global warming more likely. Further, global warming impacts will continue even after CO2 emissions are reduced since it will take time for nature to lower CO2 concentrations in the atmosphere. The impact of global warming on weather and climate change must be better understood and quantified in order to make informed decisions on CO2 emissions. Extreme points of view are well known; for instance, that climate change is not man made; to the idea that burning fossil fuels will lead to climate change with catastrophic results. What is needed are science-based conclusions that both sides can agree on. Near term economic costs of reducing CO2 emissions versus long term benefits of doing so should be quantified so tradeoffs can be judged properly.

It is apparent there will have to be a transformation in energy that reduces the amount supplied by petroleum and natural gas in the future. If present trends continue, this transformation will occur when peak world fossil fuel production capacity is reached, or international trade disruptions occur. In this case, global warming will continue, and the transformation will be much more difficult due to a lack of planning. A wiser choice is to begin a planned transformation now to avoid going through it in a crisis situation.

Options available to make transformations in energy supplies possible are conservation, nuclear, renewables (wind, hydro, photovoltaics and biomass) and alternatives to petroleum for transportation. Renewable sources of energy and alternatives to petroleum for transportation require storage of electrical energy to make them more viable. Comments on each follow.

## CONSERVATION

Conserve, conserve, conserve. Energy saved today is energy for the future. Conserving
does not just mean doing less. It also means doing things more efficiently. Significant efficiencies can be achieved in areas such as transportation, air conditioning, heating and lighting. Improved efficiencies represent the single highest impact potential with the least disruption to world economies.

## NUCLEAR

Three significant issues stand in the way of increasing nuclear energy in the future: operating safety; disposal of spent fuel; and prevention of upgrading fuel for making weapons of mass destruction. All three will require international agreements.

Nuclear fuel reserves: An estimated 200-year supply of nuclear fuel exists in known ore deposits for plants in operation today. More efficient technology is available for new plants. The possibility also exists for breeder reactors to greatly extend the expected life of known reserves. Further, uranium can be extracted from the oceans if prices grow. Energy from nuclear reactors has tremendous upside. It should be a priority to address the technical and political issues.

## HYDRO (SOLAR)

Most large projects have already been completed or are in construction. Smaller projects are feasible, and their utility and benefits can be increased by using them as storage facilities for electricity from wind and solar as discussed in the storage section.

## BIOMASS (SOLAR)

Biomass has two potential benefits in addition to a source of energy; growing biomass takes CO2 out of the atmosphere and biomass can be converted to transportation fuels to reduce reliance on petroleum. However, the CO2 and energy balance of the total biomass cycle must be positive. For example, some technical analysts suggest that ethanol as a supplement in gasoline and diesel fuels results in a zero net energy increase. Algae looks promising as a biomass energy source. Research in this as well as other biomass sources should continue.

## WIND (SOLAR)

Wind is proven technology currently in wide use in more developed countries. Significant potential exists in underdeveloped countries. The ratio of benefit to cost is very favorable.

## PHOTOVOLTIACS (SOLAR)

Photovoltaics is proven technology and in use in more developed countries with significant potential in underdeveloped countries. Benefits will increase as costs are reduced.

## SOLAR ENERGY AVAILABLE

M. King Hubbert in an American Association of Physics paper November 1981 gave the following distribution of energy influx from the sun:

| | |
|---|---|
| Direct reflection into space | 30.00% |
| Direct conversion into heat | 46.90% |
| Evaporation, precipitation, etc. | 22.90% |
| Wind, waves, convection and currents | 0.20% |
| Photosynthesis | 0.02% |

It is interesting to note that only 2 tenths of 1% of the sun's energy was converted to wind which supplied 2.5% of the world's energy in 2018. Further wind energy we capture in the form of electricity is only a minute fraction of total potential wind energy. This indicates many times the amount of energy consumed by the world is available from the sun given the ingenuity to capture it.

## STORAGE OF ELECTRICAL ENERGY

As renewable sources of energy replace fossil fuels, storage of electrical energy will be critical. Solar and wind energy are intermittent requiring storage for when they are not available. Transportation require storage of electrical energy as vehicles will not be on the grid when mobile. Three broad types for storage applications are; battery, hydro and hydrogen for use in fuel cells.

Battery: Battery storage of electrical energy has been around for a long time. To date most batteries have been for relatively small applications. Use in transportation require larger, lighter and more efficient types. Even larger (giant by comparison) batteries to store electrical energy from intermittent sources such as wind and photovoltaics are being put into service. Research should continue to make all battery installations lighter and more efficient.

Hydroelectricity complexes: A portion of electricity generated at the base of some hydroelectric dams is used to pump water back into the reservoir during periods of low demand in order to provide more electricity during periods of high demand. Indirectly electricity from intermittent sources fed into the grid is used in this practice. Essentially the complex can be used as a battery. Use of this practice enhances the utility of wind and solar sources. Further it can be an incentive to install smaller hydroelectric projects.

Hydrogen for fuel cells: Hydrogen can be recovered by electrolysis of water. Electrolysis works by applying an electrical current to water which causes the water to separate into its two components hydrogen and oxygen. Output from intermittent sources of electricity can supply the electrical current. The hydrogen can be stored in pressured vessels for later use in fuel cells. An added benefit is that fuel cells can be used in transportation.

## CONCLUSIONS

Adequate supplies of energy are available. However, there will be a transformation from relying heavily on fossil fuels to major increases in renewable sources of energy. Timing is uncertain but will be driven by one or more of three things: shortages of petroleum and/or natural gas; political unrest in supply areas; and/or concern over the impact of global warming and climate change.

Given normal world conditions there is time for an orderly transition from petroleum and natural gas to renewable sources of energy before peak production occurs within the next 25 years or so. If the transformation is driven by political unrest or concern over climate change, it could be forced at any time with significant economic fallout.

Even under the best of circumstance 25 years is not a long time. This and uncertainty relating to the last two issues suggest the move to renewables should accelerate beginning now.

Three options are available to manage the transformation:

1. Let private enterprise and market forces dictate the pace and direction
2. Empower the government to set the pace and direction
3. Combination of 1 and 2

The third option is the only viable choice. The powerful force of private industry driven by market forces must be allowed to work in this massive undertaking. The government must provide appropriate overview to assure public safety and environmental protection. Also, the government must provide incentives where sufficient near-term profit potential does not exist for needed expenditures on research and development.

Government's role is complicated due to wide and diverse national governments with different vested interests in energy production and consumption. Herculean efforts will be required to obtain international cooperation and agreements.

An active dialogue must be maintained between the public, the government and private enterprise to achieve proper balance. Hopefully, this writing has placed you in a position to participate in that dialogue.

Made in the
USA
Lexington, KY